Fire Safety in the Atomic Age

Lawrence E. Whitman

Nelson-Hall nh Chicago

Library of Congress Cataloging in Publication Data

Whitman, Lawrence E
Fire safety in the atomic age.

Bibliography: p.
Includes index.
1. Radioactive substances—Fires and fire prevention.
2. Radioactive substances—Safety measures. 3. Radiation—Safety measures. I. Title.
TH9446.R3W44 614.8′39 79-17265
ISBN 0-88229-529-2 (Cloth)
ISBN 0-88229-732-5 (Paper)

Manufactured in the United States of America

10 9 8 7 6 5 4 3 2 1

Dedication

Thirty-odd years ago Winston Churchill, speaking of the warring nations of Europe, compared the continent to a vast prison the doors of which were open. All that was lacking was the will to walk out into the sunshine of peace and freedom. The situation in which we find ourselves is very similar to that referred to by Sir Winston.

Our civilization, as we know it, has many aspects of a prison. The walls confining our society are ignorance; the roof is blind, unreasoning fear of the temporarily unknown. Outside the confines of our own making is a world far better than the one we know. It is up to us to step through the open door into that world.

Because the author firmly believes that we are only at the beginning of a higher order of life, this book is respectfully dedicated to those free spirits who are pushing back the walls of ignorance regardless of the obstacles facing them.

YOU SHALL KNOW THE TRUTH AND THE TRUTH SHALL SET YOU FREE.

John 8:32

Contents

Preface

Although radioactive materials had been known to exist for many years, they were not widely used until after World War II. Primarily, the lack of use was due to the prohibitive costs of the then known methods for extracting radioactive materials from the ores in which they were found. These high costs limited the use of these radioactive materials to a very narrow area, so the general public, including most fire safety organizations, had very little knowledge of potential radiation problems or of how to deal with these problems.

Had the use of radioactive materials remained limited, the lack of knowledge on the part of the general public and fire safety organizations would have been of minor importance. Further, if the use of radioactive materials had increased gradually as has been the case with most hazardous materials, there would have been a corresponding gradual increase in the knowledge of potential hazards and in the ability to deal with them. Unfortunately, the widespread use of radioactive materials did not come about gradually. It came suddenly under the spur of war.

On July 16, 1945, the first mushroom cloud arose over Alamogordo, New Mexico. A few weeks later Hiroshima and Nagasaki were destroyed by atomic blasts with the loss of tens of thousands of lives and hundreds of millions of dollars in property. In these three mushroom clouds the Atomic Age was born. The genie is out of the bottle and there is no way to put it back. The problem now is to learn how to live safely with the new powers granted to mankind and how to use these powers for the general good. It can and must be done.

Because of the way the Atomic Age arrived, too much thinking has been colored by fear—by the thought of mushroom clouds, the ensuing shock waves, the fireballs, and the fallouts. This is a poor approach and will lead nowhere. If we are to handle peacetime radiation hazards properly, it is necessary first to separate wartime from peacetime hazards. Once this is done, it is possible to break the problems down into their component parts. Then radiation problems become manageable.

Fire safety personnel with rare exceptions are not economists, foreign affairs experts, ecologists, or physicists; neither should they be expected to be so. They are, however, or should be, knowledgeable of fire hazards and of the means to control those hazards. If fire safety people do their part and other experts do theirs, there is no reason why radioactive materials cannot be safely used in almost every type of activity, from energy production to medicine to manufacturing.

The purpose of this book is to present in layman's language a basic picture of the types of radiation hazards and the methods used to deal with these hazards. In the process we will present certain fundamental facts and endeavor to clear away many common misconceptions regarding radiation hazards.

Far too much of what appears in newspapers and

magazines, on television and radio, and on the rostrum is presented for sensational impact and implies that radiation is far too complex for the average individual to understand. This is the gospel of fear and is simply not true. A common-sense approach to the subject will show nonexpert individuals that they can do much to protect themselves and others from any radiation hazards they are likely to encounter. The end result of a common-sense approach will be to place radiation hazards in the proper perspective in relation to the other hazards of everyday life.

Radiation should not be feared; neither should it be ignored. It can and should be understood. Then the enormous powers available from radioactive materials can be harnessed and put to work. That is our job.

Acknowledgments

No book of this kind is ever the product of one individual. It is rather a team effort in which many persons have a part. Some play a direct role, while others participate through encouragement and moral support. All are important and their activities have been deeply appreciated by the author. A number of individuals made major contributions which deserve special recognition. These friends are:

Charles C. Mosby, FF III, Lynchburg, Virginia, Fire Department, who provided a great deal of assistance in running down sources of information and who also did some of the artwork required for the illustrations.

Charles A. Droog, Commander, and Ted R. Bratton, FF II, both members of the Lynchburg Fire Department, who did the bulk of the illustrations used in this book.

Joe E. Johnson, FF II, Lynchburg Fire Department, who assisted in the preparation of illustrations.

The late Edgar H. Dewell, Associate Professor of Nuclear Technology, Central Virginia Community College, whose advice on a number of phases of radiation physics was very helpful.

The late Charles Van Valkenburg, Captain, New York City Fire Department (Ret.), who patiently waded through reams of typed material to provide a fire service review.

The officers and men of the Lynchburg Fire Department, whose encouragement and interest have been of tremendous help in carrying this project through to completion.

Credits for Illustrations

1. Permission to use the following figures has been granted by the United States Energy Research and Development Administration: Figures 1, 2, 3, 4, 5, 6, 7, 8, 16, 17, 19, 20, 21, 22, 23, 24, 25, 26, 27, 28, 48, 49, 53, 54, and 62.
2. Permission to use the following figures has been granted by the United States Department of Transportation: Figures 33, 34, 35, 36, 37, 38, 39, and 41.
3. Figure 42 is made up of four original drawings of equipment used by the Lynchburg, Virginia, Fire Department. Since the same types of equipment are used in many locations, there are undoubtedly similar drawings elsewhere.
4. The following figures were prepared by Commander Charles A. Droog, FF III Charles C. Mosby, FF II Ted R. Bratton, FF II Joe E. Johnson, all of the Lynchburg Fire Department. Figures 9, 10, 11, 12, 13, 14, 15, 18, 29, 30, 31, 32, 40, 43, 44, 45, 46, 47, 50, 51, 52, 55, 56, 57, 58, 59, 60, 61, 63, 64, 65.

1

Introduction

Ionizing radiation, what it is and what it can and cannot do, is little understood by the average individual. What scant information the layman has is largely gained from: (1) sensational but not particularly informative articles written about some unique real or imagined hazard and (2) technical articles written in language intelligible only to the trade. Such contacts, plus the real fear generated by the known capabilities of atomic weapons and the instinctive "shying away" from an unknown that is colorless, odorless, and tasteless and can be neither heard nor felt make many people want to "leave it to the experts."

Because the layman has not been informed in lay language of the real hazards of radiation and what he or she can do about them, two unfortunate tendencies have developed. First, many individuals ignore the subject completely and forget about the potential hazard to themselves and to others. For example, in a firefighting situation it is entirely possible to carry out necessary operations without undue hazard to personnel. It is also possible by ignoring the situation to expose personnel unnecessarily to radiation hazards. Second, many individuals develop an unreasoning fear of all radiation and consequently oppose any project involving radioactive materials even though no real hazard exists. For example, two nuclear power plants are proposed: Plant No. 1 would straddle an earthquake fault and contain large amounts of combustibles in its construction and Plant No. 2 would be located in a stable area and contain virtually no combustibles. The fearful individual would automatically oppose both plants, while the informed individual would oppose only Plant No. 1.

The purpose of this book is to place radiation hazards in their proper perspective relative to other contemporary problems. To do this, radiation hazards will be discussed in layman's language. There is no need for the average individual to become a scientist in order to deal with the radiation hazards he or she is likely to meet. What is needed is a working knowledge of the kinds of hazards and how to avoid these hazards or deal with them so there will be no unnecessary exposure.

The Layman's Approach to Radiation

In presenting the radiation hazard problem, we will generally use the working concept approach, which is a natural and easy way of learning a new subject. Precise definitions, which for the most part are of real interest only to scientists, engineers, laboratory technicians, and the like, will be used only when their use is essential to understanding a particular problem. This approach makes it possible to develop some ideas of physical phenomena that are accurate enough to accomplish the goal of radiation safety. The results may not be as accurate as those required by a physicist, but they will meet our needs.

The Atomic Age has given mankind the ability to improve the quality of life greatly. Fallout from radiation research has already given us tools to cure some malignant diseases, improve the preservation of food, provide better quality control in industry, produce energy, and develop new and useful products. Despite these great benefits, one fact makes it necessary for anyone who is likely to come in contact with radioactive materials to have a basic understanding of radiation hazards. That fact is: All radioactive materials emit energy that has the power to damage living tissue.

Defining Radiation Control Systems

At the beginning of any layman's course on radiation it is appropriate to note that radiation hazards are not normally avoided by any single safety measure. Rather, safety is assured by setting up and maintaining a control system suitable to the hazard involved. Far too many people think of a control system as being an assembly of devices arranged to do certain things under either automatic or manual control. Such an assumption is false and can easily create an illusion of safety that isn't really there. Any effective control system for radiation hazards requires knowledgeable personnel and proper construction as well as automatic and/or manual control. Of these three keys the need for properly trained personnel has unfortunately received the least emphasis.

No matter how well designed and maintained monitoring devices and control equipment are, they are no better than the people using them. In the hands of well-informed and competent personnel control devices and allied equipment can be invaluable tools in locating and controlling radiation hazards with minimum risk to life and property. By contrast, the same equipment in the hands of ill-trained and incompetent people can readily increase the hazard through: (1) lack of understanding of what the devices and equipment are doing and indicating and (2) misuse of nonautomatic controls. Never lose sight of these facts:

1. Control devices and equipment are tools designed to assist individuals responsible for the control of radiation hazards. They are tools and nothing more.

2. Protective construction is intended to provide a shield between personnel and radiation hazards. To be effective, however, the shield must be used properly.

3. Effective use of control devices and equipment requires proper training of responsible personnel.

4. Personnel assigned supervision over radiation hazards must have sound judgment.

5. As previously noted, radiation is colorless, odorless, and tasteless and cannot be heard or felt. Therefore, unless people know what they are doing and why, they can very quickly get into serious trouble through unnecessary exposure to a silent killer.

From these axioms it is clear that a good radiation control system is one in which qualified personnel of sound judgment have been provided with all necessary information relative to the hazards involved and are then given the tools to do the job. These tools include not only radiation monitoring and control equipment but also essential fire-fighting and salvage equipment and last, but not least, the authority to eliminate unacceptable hazards prior to the development of a fire situation. When all these factors are properly meshed, an effective fire safety radiation control program can exist.

Requirements for an Effective Control System

Assuming qualified and adequate personnel, there is a logical order to be followed in developing a good radiation control system. Since most students using this book probably will be members or prospective members of fire departments or industrial fire organizations, it is appropriate to detail the essential steps from the standpoint of the fire service. Other organizations and individuals should use the same basic approach, modifying details as required to meet their particular situations. Basic steps are as follows:

1. *The Overall Radiation Problem.* Without factual knowledge of the locations, types, and amounts of radioactive materials and the types of radiation hazards involved, the fire service (or for that matter any other fire safety organization or individual) is flying blind. The fire service also must have essential data on locations and types of nuclear reactors, particle accelerators, and any other radiation machinery that may be present in their area of responsibility.

2. *Judgment.* In order for personnel to exercise sound judgment, information relative to past radiation fire incidents is vitally important. Such data, properly used, are invaluable in programming for the prevention of similar fire incidents and for minimizing the risk to personnel and property should a fire occur.

3. *Acceptable Risk.* Since the Atomic Age is well underway, steadily increasing use of radioactive materials must be anticipated. A clear understanding of what constitutes an acceptable risk is therefore essential. The degree of risk that is acceptable will of course vary with the importance of the proposed operation. For example, more risk is justified for national security than could ever be justified in a nonvital operation.

4. *Types of Radiation Hazards and Neutron Bombardment.* The next step is to review the various types of radiation hazards and/or neutron bombardment hazards to determine how to deal with the hazards that may be present.

5. Contamination. In a fire involving radioactive materials, a certain amount of radioactive contamination is inevitable. Much of the potential contamination in any given situation can be avoided by careful preplanning, but the fire service must deal with the remainder.

6. Instrumentation. Many devices designed to record radiation levels have been developed. Some are interchangeable and some are not. Not all devices will record all types of radiation. Instruments used must be suitable for the hazard involved. Incorrect instrumentation is not only useless, but it may permit a major radiation hazard to pass unnoticed.

7. Control Equipment. Safety controls and instructions for their use are the responsibility of nuclear engineering designers. Where controls are provided, personnel must be carefully instructed not only in their use but also in actions to be taken in the event of an equipment malfunction.

8. Radiation Physics. Necessary to a proper understanding of radiation hazards is a layman's introduction to radiation physics. Three vitally important facts are:

a. Radioactive emissions from a decaying atom are not necessarily constant, although many are. Some atoms may emit alpha rays at one stage, beta rays at another, and gamma rays at still another. Such materials may also give off all three types of radiation at the same time. Operating personnel need to know what type of radiation may be anticipated.

b. Personnel connected with nuclear energy must know the difference between fission and fusion.

c. Personnel need to know how materials can be made radioactive.

9. Transportation. Specially designed containers and packaging materials are required for transporting hazardous amounts of radioactive materials. Requirements also mandate identifying materials and the degree of hazard. For these safeguards to be fully effective it is essential that personnel with fire safety responsibility understand the packaging system and know how to read identifying markers.

10. Large-Scale Operations. In this area fall nuclear power plants, particle accelerators, and other radiation machinery. Fire safety personnel who might be involved in such activities need an introduction to operations and an awareness of the potential dangers from the presence of excessive amounts of combustibles.

Satisfactory completion of items 1 through 10 above makes it possible to preplan not only normal day-to-day activities but also actions to be taken in the event of fire or other accident. Preplanning would also take into account how and where to secure assistance from radiological specialists.

History of the Atomic Age

No introduction to radiation fire safety is complete without a brief outline of how the Atomic Age came into being. The time frame containing the first discoveries, present knowledge, and projections for the immediate future is incredibly short. It is not surprising, therefore, that so many misconceptions and so much misinformation is floating around among the general public, including many who directly or indirectly are or will be dealing with ionizing radiation. Because of the potential for great benefits or for great harm (if misused), it is essential that the next chapter of history bring the average individual into the picture. Otherwise, progress will stall and unnecessary disasters might result.

Mankind has been speculating on the nature of matter since antiquity. A little of the thinking was along logical lines, but more was simply a series of wild guesses. Little by little, however, the body of factual knowledge increased, until by the end of the nineteenth century real progress became possible.

It is perhaps in order to say that modern atomic science started with ancient Greek philosophers who conceived the idea of the atom. Later this idea was expanded to include molecules, but it was still unknown that the atom was divisible and that so-called solid matter is mostly empty space. Then in 1896 Antoine Henri Becquerel discovered radioactivity and the Atomic Age began.

The discovery of radioactivity and the fact that uranium spontaneously decayed to radium and radium to lead paved the way for development of the basic theory of atomic structure. This theory was quickly proven by the discovery of the electron by J. J. Thomson in 1897 and of the proton by Ernest Rutherford in 1911. Had discovery stopped there, the uses of radiation would have been very limited due to the few types of naturally radioactive material and the cost of extracting the material from ores. Two items remained to complete the picture.

In the early 1900s Albert Einstein propounded a new theory of the relationship of mass and energy. This theory, long since proven, gave evidence that tremendous energy was locked up in the atom. This energy, released, could serve many purposes. The only thing now lacking was the means to unlock the atom. This lack was soon overcome.

Scientists had long noted that atoms of the same element did not all necessarily have the same atomic weight. Research was conducted to determine the reason, and in 1932 James Chadwick discovered the neutron, a particle with no electrical charge. Differing weights for atoms of the same element were found to be due to variations in the number of neutrons in the nucleus.

Following the discovery of the neutron it was found that both free neutrons and gamma rays could be used to split atom nuclei, releasing tremendous quantities of energy. Splitting the atom inevitably led to atomic weapons and nuclear energy equipment such as reactors. From the latter have come a new energy source and the ability to create many new and useful radioactive elements and isotopes. There is little question that many more valuable uses can be found for ionizing radiation, provided it is remembered—again—that *all radioactive materials emit energy that has the power to damage living tissue.*

2

Beneficial Uses of Radiation

Although some hazard is, admittedly, involved in the use of radioactive materials, radiation also can be put to a great many beneficial and potentially beneficial uses. Since it is possible to control radiation hazards within acceptable limits, it is desirable to examine briefly some of the benefits that can be derived from the proper use of radioactive materials.

Tracers

Radioactive materials send out signals that can be detected by electrical or chemical means. This property makes it possible to trace biological processes, control toxic residue materials, gain information that will improve agricultural processes, and improve many industrial processes. Examples of these benefits are:

1. Diagnostic Medicine. The fact that certain elements on entering the body go directly to a specific organ or organs has been known for quite some time. For example, nearly all iodine entering the body goes directly to the thyroid gland. Knowing this, a surgeon who suspects thyroid problems will arrange to have a small quantity of radioactive iodine introduced into the patient. Then it becomes possible, by recording the signals from the radioactive iodine, to determine how well the thyroid is functioning. It is also possible to determine whether or not surgery may be needed and what other presurgery tests may be required. The radioactive iodine does nothing from a curative standpoint, and its quantity of radiation is too small to do any harm. It is simply a diagnostic tool. Many other types of radioactive isotopes are also used to diagnose organ malfunctions and other bodily ailments for both people and valuable animals.

2. Poison Control. The agricultural industry uses various types of insecticides, herbicides, and fungicides. These poisonous materials normally are safely removed, but the possibility exists that in some phases of agricultural processes toxic residues may remain. Radioactive isotopes are extremely useful in checking for this possibility when the presence of toxic residues is suspected. This can be of particular importance when the pesticide is absorbed into the plant structure.

3. Fertilizer. Fertilizers may be applied to plants in many ways. Some are normally absorbed through the roots, some through the leaves, and others by either method. Efficiency of application is affected by the type of soil and by the weather pattern. Through the use of radioactive isotopes it can be determined what types of fertilizer will give the best results for any given set of conditions.

4. Food Production. The most efficient way to produce the maximum amount of meat, poultry, and dairy products is by what is essentially an assembly line process. At present a great deal (though not all) of the feeding is done at random. Although a balanced diet is the goal, random methods inevitably involve considerable waste and occasionally improper nutrition. Through the use of radioactive isotopes it is possible to get far better control of livestock feeds. This process results in more and better products, a factor of considerable importance in a world where the human population is exploding.

5. *Other Biological and Botanical Processes.* In addition to the processes noted above, research is being done on other phases of biology and botany.

6. *Industrial Processes.* The use of tracers has been invaluable in many phases of industry, including the construction industry. Of the many uses, the following are representative:

a. Gauging. Many metalworking industries, particularly plate mills and wire mills, roll metals at high speeds to exact thicknesses or diameters. Since a 1 percent error would mean the loss to the manufacturer of the cost of one ton of metal for every 100 tons rolled (not to mention the cost of rerolling), there is an obvious need for devices that will intermittently or continuously monitor the metal being rolled. Radioactive isotopes in thickness gauges permit such monitoring, and their radiation level is well within the permissible range for personnel. There is no other practical means for such monitoring without drastically curtailing permissible line speeds with consequent loss of productive capacity. Such gauges are also used to measure thicknesses of cloth, paper, and so on.

b. Metal Defects. Quite often it is necessary to know whether or not any defects exist in a key structural element or important piping that might cause the element to fail at a later date. The need to know is particularly important for large pipelines that run long distances and are not or cannot be inspected visually to detect flaws. The Alaska Pipeline, transcontinental gas and oil lines, and many high-pressure and chemical pipelines used for other purposes are examples. A combination of a radioactive isotope inserted into the pipe and X-ray film outside the weld give an X-ray picture of the strength of the weld. Such checks plus allied valve and fitting checks in the factory will, if properly done, assure a sound installation. Again, the radiation level is too low to be a hazard to personnel.

c. Wear. A substantial number of industrial operations depend on the continuous, satisfactory performance of a key piece of equipment. Uneven or too much wear on moving parts leads to poor performance. To check this, the wearing surface of the moving part has traces of a radioactive isotope in its makeup. As the part wears, the radioactive isotope will gradually be detected in the lubricating system.

d. Transportation of Liquid. Modern industry frequently requires the transporting of large amounts of liquids, some of them flammable, over long distances. Much of this transportation is by pipeline. Because the cost of pipelines is high, the possibility of using a single line for more than one product was investigated. Many liquids are not miscible; some will mix only slightly; and others can be kept apart by a seal (called an interface). Since it was definitely possible to keep different products, or for that matter different shipments, apart by a simple interface, it has become standard practice to use single lines for multiple products and/or multiple shipments. At first there was inevitably appreciable product wastage due to the difficulty of pinpointing the exact location of the interface. Now it is common practice to introduce a small amount of a radioactive isotope at the interface. With radiation sensors it then becomes possible to locate the interface exactly and eliminate the waste, saving time and the cost of disposing of the waste liquids. This use of tracers has been of great value to the oil industry.

Electrical Conductivity

Radioisotopes have the property of making the surrounding air electrically conductive. This property has been of considerable importance in many manufacturing processes, as it provides an easy and practical means for removing static charges. Nonconducting materials moving over rollers or involved in spreading operations frequently generate substantial amounts of static electricity that cannot be drained off by ordinary methods of grounding and bonding. For such situations, to avoid the possibility of fire from static sparks, ionization static eliminators provide an important means to eliminate the static. Depending on the nature of the operation these static collectors can be of several types, one of which uses radioactive isotopes. This last means is of particular importance where highly volatile flammables are used (for example, rubber cement spreading). Units function by neutralizing static charges. Normally, radioactive elements used are alpha emitters.

Tissue Destruction

A truism that must be regularly repeated is that radioactive materials emit energy that can destroy living tissue. In addition to the potential hazard, possible benefits can be derived from this property. These benefits are substantial.

1. Therapeutic Medicine. At present, the principal use of radiation therapy is in the treatment of cancer. For nonmedical personnel, the best description of cancer is a group of cells whose growth rate has run wild. If continued, this wild growth will destroy the body's ability to function and result in death. Radioactive materials such as radium and more frequently cobalt 60 have been used to kill off cancerous growths. Many of these efforts have been successful.

2. *Sterilization.* A problem in the storage of foods and in the manufacture and storage of drugs is the prevention of spoilage, particularly if long shelf-life is desired. Both sterilization and pasteurization can be accomplished through the use of radiation or heat. For some types of food and drugs, however, only the radiation method is practical. For example:

a. *Foods.* For the long-range storage of food products other than canned goods, the foods must be sealed in moistureproof paper and cartons that will not permit contact with outside air. Heat sufficient to sterilize or pasteurize the foods would in most instances destroy the vital seal. Instead, therefore, the products are subjected to sufficient radiation to kill all the living organisms (sterilization) or most of them (pasteurization). As shown later, this process does not make the irradiated products radioactive.

b. *Drugs.* In the manufacture of drugs it is often necessary to assure absolute sterility. In many instances the heat required for sterilizing would damage the drug, so radiation is used. Again, the process does not make the drugs radioactive.

Atomic Electricity

Electricity can be produced directly from the energy released from radioactive materials, but this electricity should not be confused with that produced by indirect means by nuclear reactors. Direct atomic electricity is produced in very small amounts, usually at relatively low voltage. The principal use at present is for atomic batteries, which are used to power small devices requiring a dependable, constant, and noninterruptible power source (an example is pacemakers).

Fission Energy

The beneficial uses of radiation discussed so far involved the energy given off by the natural decay of radioactive materials. Some common uses of these types of radiation are illustrated in Figures 1 through 8. Radiation levels used are generally low and easily controllable from the standpoint of personnel safety. There are, however, other ways in which energy can be obtained in great quantity from the atom.

Some materials are able to emit bits of matter that, on striking other atoms of the same material, cause them to break open. The broken atoms emit more pieces of matter, which in turn break open more atoms. Each time an atom is broken open some of the energy that had been binding the atom together is released. This energy can be harnessed and put to work. It is this property that makes atom-splitting useful. Energy released by this process is known as fission energy. It should not be confused with normal radiation for two reasons: (1) most radioactive materials are not fissionable and (2) the energy release is vastly greater than that obtained from the natural decay of radioactive materials.

The atomic fission process, which will be described in more detail in Chapter 11, provides mankind with many new powers, among which are production of electricity, creation of new radioactive elements, and breeder reactors.

1. *A New Method of Producing Electricity*. The energy released through one of several methods is converted to electricity. Often this conversion can be accomplished at substantially less cost than is possible with conven-

FIGURE 1

RADIOACTIVE PHOSPHORUS-P 32

FOR STUDY OF PHOSPHATE FERTILIZER UPTAKE

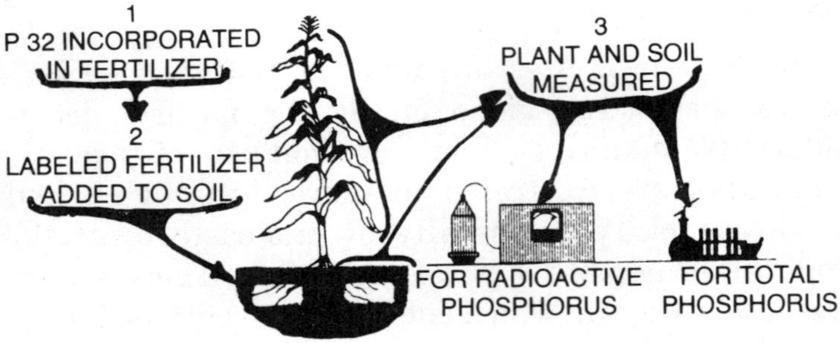

SHOWS:

1—FIXATION OF PHOSPHORUS BY SOIL
2—PHOSPHORUS UPTAKE BY PLANT
3—PROPER TYPE AND PLACEMENT OF FERTILIZER
4—EFFICIENCY OF FERTILIZER

FIGURE 2

RADIOACTIVE ISOTOPES

FOR TRACING OIL FLOW IN PIPELINES

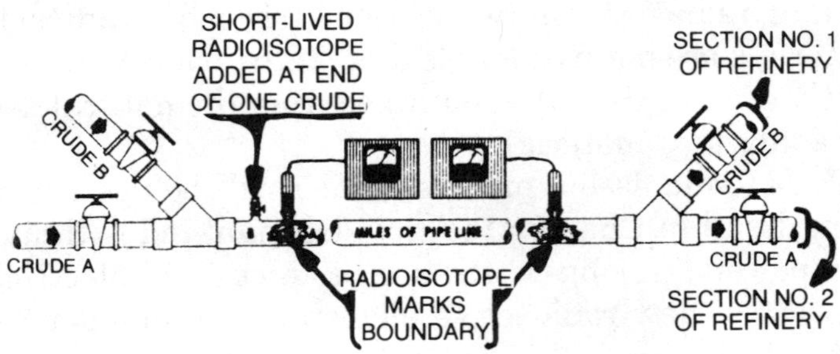

ADVANTAGES:

1—RADIOACTIVE "MARKER" CAN BE TRACED OVER LONG DISTANCE
2—"MARKER" SPREADS TO ONLY SMALL OIL VOLUME
3—PERMITS SEPARATION OF CRUDES WITH MINIMUM OF LOSS
4—METHOD QUICK AND REQUIRES NO SAMPLING

FIGURE 3

RADIOACTIVE COBALT—Co60

FOR RADIOGRAPHY TESTING

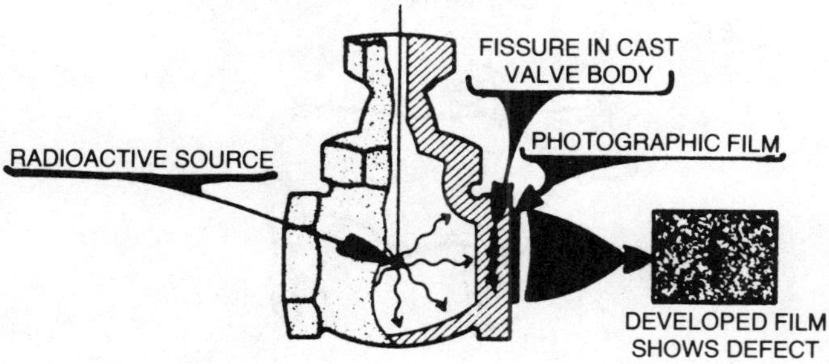

ADVANTAGES:

1—VERSATILE AND RELIABLE INSPECTION
2—INSPECTION MADE WITHOUT DISMANTLING
3—SOURCES OF DESIRED SHAPE AND SIZE
4—VERY HIGH ACTIVITY SOURCES AVAILABLE AT LOW COST

FIGURE 4

RADIOACTIVE IRON—Fe59

FOR FRICTION AND LUBRICATION STUDIES

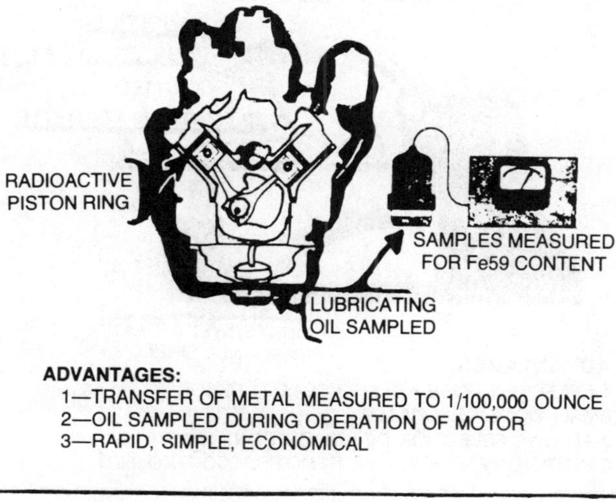

ADVANTAGES:

1—TRANSFER OF METAL MEASURED TO 1/100,000 OUNCE
2—OIL SAMPLED DURING OPERATION OF MOTOR
3—RAPID, SIMPLE, ECONOMICAL

FIGURE 5

TRACING WITH RADIOISOTOPES
USING
CARBON 14

1 MILLIGRAM OF C14 EMITS 200,000,000 RADIOACTIVE SIGNALS/SEC.
1 MG C 14 IN A DRUG, VITAMIN OR HORMONE INJECTED INTO A 1,000 POUND COW CAN STILL BE MEASURED IN 10 MG. AMOUNTS OF BLOOD, MILK, OR TISSUE.

FIGURE 6

ROTATIONAL TELETHERAPY
USING
Co 60 GAMMA RAYS
(FRANCIS DELAFIELD HOSPITAL—NYC)

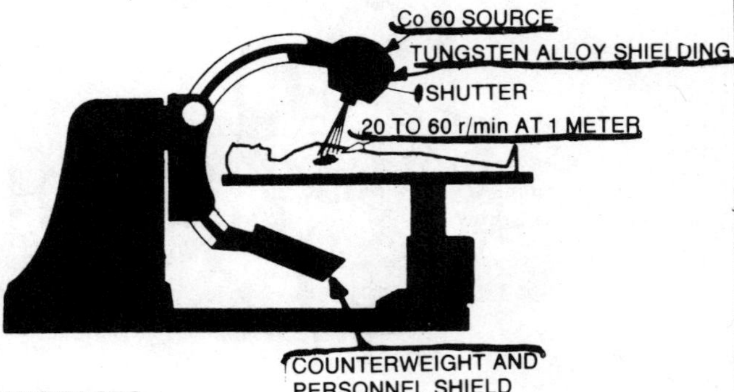

ADVANTAGES:
1-EFFECTIVE DOSE AT DEEP-SEATED TUMOR—SMALL DOSE AT SURFACE
2-ALLOWS SELECTION OF IRRADIATION PATTERNS
3-ROTATION AND SHUTTER REMOTELY CONTROLLED

FIGURE 7

RADIOACTIVE IODINE—1·131

FOR DIAGNOSING AND TREATING THYROID GLAND DISORDERS

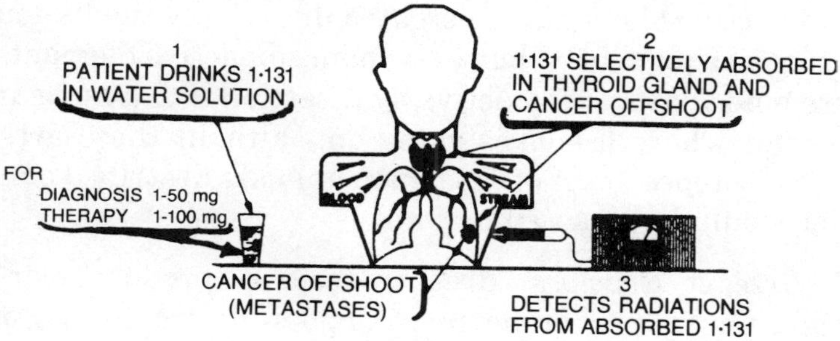

MEDICAL ACTION

1—DIAGNOSIS AND TREATMENT OF HYPERTHYROIDISM
2—LOCATION OF THYROID CANCER OFFSHOOTS (METASTASES)
3—TREATMENT OF THYROID CANCER AND METASTASES

FIGURE 8

RADIOACTIVE SOURCE

FOR GAUGING THICKNESS

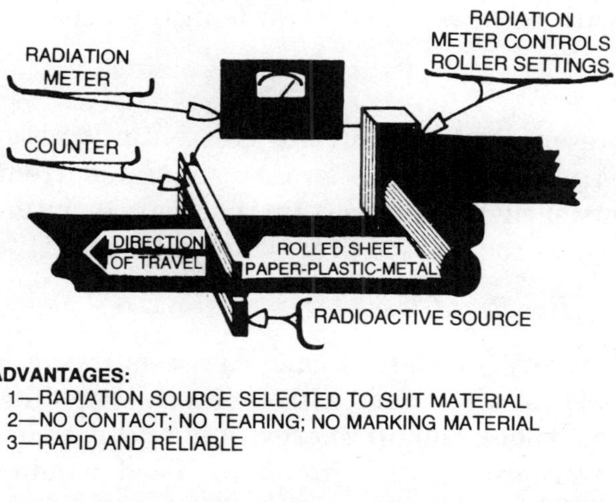

ADVANTAGES:

1—RADIATION SOURCE SELECTED TO SUIT MATERIAL
2—NO CONTACT; NO TEARING; NO MARKING MATERIAL
3—RAPID AND RELIABLE

tional fuels. Contrary to the belief of many people, it is possible to design and operate such plants safely, as proven by actual operating records.

2. Creation of New Radioactive Elements. New radioactive elements are created directly by the fission process. Frequently, however, nonradioactive elements are made partly radioactive by insertion into a nuclear reactor where fission is going on. Without these artificial isotopes, most of the beneficial uses described earlier would be impossible.

3. Breeder Reactors. Breeder reactors produce more fuel than they consume, making possible greater use of nuclear power where needed.

There are and undoubtedly will continue to be persons who object to the use of fission energy. These people are for the most part psychological descendants of those who objected to iron ships because iron sinks or who thought rail traffic at thirty miles an hour was impossible because people couldn't breathe at that speed or who ridiculed the idea of stopping heavy trains and trucks with air brakes. There are admittedly hazards, but these hazards can be dealt with as have other hazards in the past.

Fusion

At present the fusion process is essentially restricted to weaponry. Because the process liberates tremendous amounts of energy, it is certain that ways eventually will be found to control the process for a beneficial use.

Areas of Beneficial Use

Fire safety in the Atomic Age requires a general knowledge of where radioactive materials are likely to be found or where radiant energy is likely to be used. The following representative groups are listed to indicate how widespread such uses are.

1. Industry
   Metallurgical processing
   Wire mills
   Chemical processing
   Measuring and thickness gauges
   Food irradiation
   Mining
   Radiographic examination: construction, foundries, etc.
   Corrosion control
   Lubrication efficiency control
   Static control
2. Institutions
   Hospitals
   Clinics
   Nursing homes
3. Laboratories
   Biological
   Chemical
   General industrial
4. Utilities (nuclear power plants)
5. General
   Locating bulk movements in pipelines
   Particle accelerators
   Warehouses
   Aerospace operations

Obviously, ionizing radiation can be put to work to benefit mankind in many ways. The outstanding safety record to date is proof that we can live with radiation.

3

The Radiation Hazard Problem

All human activity involves some degree of risk and the use of radioactive materials is no exception. The risk may be physical or psychological, but it will be there. It is normal, therefore, for people consciously or unconsciously to weigh the benefits to be gained from a course of action against the possible risk. When the good to be accomplished outweighs the danger, the risk is usually accepted. Efforts are generally made to reduce the risk as much as possible, but there is no guarantee of 100 percent safety.

Even though the necessity of accepting reasonable risks in order to gain worthwhile benefits is generally accepted, many people fail to apply this principle to radiation hazards. In part their fears stem from the known capability of atomic weapons to create widespread destruction. It is probable, however, that most of their fears are generated by sensational reporting which can only be counteracted by providing the information needed for understanding. Until they have adequate knowledge, some people will continue to demand absolute safety from all radiation. Such activities place unnecessary roadblocks in the way of putting radiation to work for mankind.

Seen in the proper light, radiation hazards should be neither feared nor ignored. Rather, they should be understood so that they can be dealt with properly. Then radiation can become a useful tool whose hazards can be contained within acceptable limits.

Radiation Axioms

The first step in controlling radiation hazards is to gain a clear understanding of certain characteristics of radioactive materials and certain principles regarding their use. Of primary importance are the following:

1. All radioactive materials emit energy that can damage living tissue.

2. All radiation, if received in sufficient quantity, will damage the body.

3. Within definite limits, radiation damage to the body can be repaired so there is no apparent effect.

4. The amount and type of radiation produced by any activity involving radioactive materials can be measured by existing instrumentation.

5. Where acceptable limits of radiation can be maintained, exposure of personnel to radiation in the course of performing useful work is acceptable.

6. The degree of permissible exposure must be directly related to the importance of the work to be done. Exposure limits should be carefully set, based on such considerations as:

   a. The nature of the work being performed: (1) personnel rescue, (2) critical operations, or (3) property preservation (including routine operations).

b. The nature of the exposure: (1) regular, (2) intermittent, or (3) one time only.

7. Radiation exposures from new activities in which radioactive materials are used vary only in degree from radiation experienced from birth by all living organisms and by mankind from routinely accepted X-rays and the like.

Natural Radiation Exposure

All living things are exposed to cosmic radiation from outer space. However, the amount of exposure an individual receives during a lifetime depends on location: below ground cosmic radiation is negligible; at sea level it is slight; but as altitude increases so does the level of cosmic radiation. There is no evidence, however, that the level of cosmic radiation at any altitude poses a threat to mankind.

Aside from cosmic radiation, minute quantities of radioactive materials are present in the earth's crust and in many of our water supplies. It is also likely that very minute amounts of radioactive materials enter the body from the simple processes of eating and drinking. Again, there is no evidence that the level of radiation from these sources is in any way harmful to life. It should be evident, however, that from the very beginning, all life has been exposed to inescapable natural radiation.

Medical and Dental Radiation

Over and above the natural background radiation that is always with us, much of the population in technically developed nations receives radiation from accepted medical and dental procedures. X-rays can be of great assistance to medical and dental surgeons in minimizing risks to patients. Cobalt 60 radiation to destroy cancer cells is another example of useful and accepted radiation in medicine. By contrast, unnecessary exposure to radiation has occurred in the unwise use of X-rays by poorly trained or incompletely informed personnel. Examples are public chest X-rays for tuberculosis with no regard for how many X-rays an individual has had and fitting shoes by the use of X-rays, a process that can be replaced by other equally effective means. Fortunately, such unnecessary radiation exposure is rapidly being eliminated. Whenever radiation for medical or dental purposes is proposed, its benefits must be weighed against the possible harmful effects. Procedures in which benefits outweigh the possibility for harm should be retained. If the potential for harm outweighs possible benefits, the specific procedure should be modified. Such instances, however, should never be used as arguments for eliminating all radiation exposure.

Analysis of Common Hazards

When use of a radioactive material is proposed in industry, in agriculture, or elsewhere, the same reasoning should be applied as for other hazards. What benefits can be gained by the proposed process? What are the potential harmful effects? What means are available for controlling the hazard and how effective are they likely to be? When all the factors are added up, do the benefits outweigh the risk? If so, implementing the process is in order. If not, the project should be abandoned or shelved until better controls are available. Here are examples:

1. *The Automobile.* Without automotive vehicles, a substantial part of the ability to move vitally needed goods, to provide emergency services, and to get employees to their places of work would be lost. Automotive vehicles also cause thousands of deaths and injuries yearly. However, when benefits and hazards are weighed against each other, the good outweighs the bad. Accordingly, society decides to keep the automobile.

2. *Cobalt 60 Radiation.* Cobalt 60 radiation kills cancer cells. It can also damage nearby healthy cells. When successful cancer treatment is weighed against occasional adverse side effects, the benefits are found to more than offset any harmful effects. Therefore, this method of cancer treatment is considered acceptable.

3. *Immunization.* Inoculating or vaccinating people has virtually wiped out diphtheria, smallpox, polio, and other major diseases. A few people suffer side effects from immunization and some may die, but compared to the former wholesale casualties from epidemics, the risk is worth taking.

4. *Radiation Hazards in Industry.* As previously noted, the use of radioactive materials in industry is widespread. In general, the amount of radiation exposure is so low that the potential for harmful effects is negligible and the safeguards are, for all practical purposes, completely effective. These small-scale uses of radiation are with few exceptions accepted.

Major Potential Hazard Areas

While the safety record of operations involving radiation is substantially superior to that of other types of hazardous operations, the general public and the fire service are concerned about the hazard potential of large-scale activities. Among the most important of these concerns, each of which will be detailed in later chapters, are:

1. *Contamination (Chapter 8).* As with many industrial, agricultural, and energy-producing activities, many operations involving radioactive materials produce waste material that is in part radioactive and must be disposed of safely. In other instances radioactive materials might be involved in a fire or natural

disaster. Preventive measures to control such possible contamination must be taken in advance.

2. Transportation (Chapter 9). Since radioactive materials are widely used, it is obvious that they must be transported frequently in all areas of the country—from mining to processing and manufacturing plants to points of use, and waste materials from points of use to disposal points. Hazards may be great or small and transportation may be by air, rail, road, or water. The method of transportation is not, however, the critical factor. Knowing or not knowing the nature of the hazard and how to deal with it is the critical factor for the shipper, the transportation agencies, and emergency forces responding to an accident.

3. Nuclear Power Plants (Chapter 12). Nuclear reactors are types of equipment in which a self-supporting chain reaction can be produced and controlled. Reactors may be used to produce energy, create radioactive isotopes, or produce more radioactive materials than are consumed. Because of the amounts of radioactive materials that are normally present, it is essential that great care be taken in the selection of sites, building construction, operational safeguards, disposal of waste and surplus materials, and so on.

4. Radiation Machinery (Chapter 13). Radiation machinery includes equipment varying from X-ray machines to very large particle accelerators. Many large units are located in rural areas. Of particular concern in setting up safe installation and operation are: (*a*) the possible presence of large amounts of flammable material, (*b*) the presence of high-energy electrical equipment, (*c*) high radiation levels during operating periods, and (*d*) location of radioactive targets. All these factors can be controlled.

5. Nuclear Weapons (Chapter 14). Extreme care has

been devoted to the development of nuclear weapons, and, at this writing, no inadvertent or unintentional nuclear explosion has occurred. While the possibility of a weapon being involved in an accident cannot be entirely ruled out, the accidental detonation of the nuclear material is virtually impossible. What remains, therefore, is the hazard from radioactive materials being dispersed by an accident, by fire following an accident, or by detonation of the high explosives that may be present.

Effects of Overexposure to Radiation

Another phase of the radiation hazard problem is the lack of understanding by the general public and many elements of the fire service as to the effects of various levels of radiation exposure. These effects are shown in many ways.

1. *Radiation Sickness.* Radiation sickness is produced by a massive overdose of penetrating external radiation. Depending on the amount of radiation and the duration of exposure, symptoms include nausea, vomiting, diarrhea, and hemorrhaging. In extreme cases death will result. Because of the carefully designed safety systems in areas of high radiation, the likelihood of any individual receiving such an overdose is infinitesimal.

2. *Radiation Injury.* Whereas radiation sickness affects the whole body, radiation injuries result from overdoses of penetrating external radiation to a part or parts of the body. Injuries are therefore localized. Burns, skin lesions, and hair loss are the more common types of injuries. Hands and forearms are the areas most frequently affected. Genetic damage can also be caused by localized overdoses of penetrating external radiation.

3. Radioactive Poisoning. Radioactive poisoning is sickness resulting from excessive amounts of certain radioactive materials entering the body. Various types of cancer, anemia, and some blood diseases can be caused by this type of poisoning.

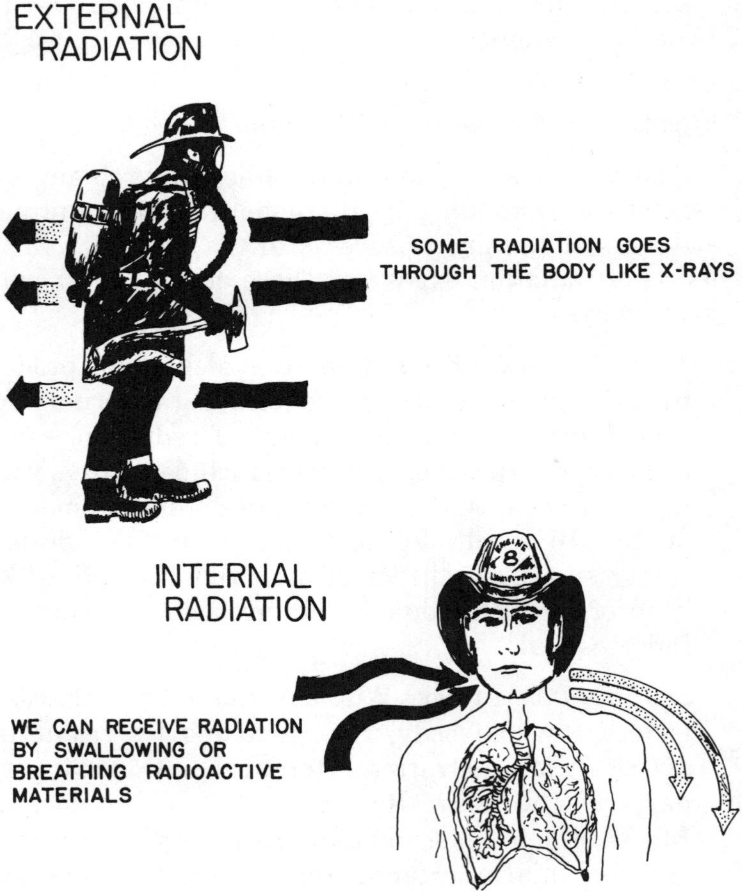

FIGURE 9

Overcoming the Knowledge Barrier

Last but not least of the factors involved in the radiation hazard problem is overcoming the psychological and knowledge barriers that stand in the way of using radioactive materials wisely and well for the common good. In most instances breaking the knowledge barrier will at the same time break the psychological barrier. The most practical approach is to:

1. Recognize that the radiation hazard problem is twofold, including radiation from sources outside the body and radiation from materials ingested into the body (see Figure 9). As we proceed it will become evident that precautions necessary for one type of hazard will often differ substantially from those required for protection against the second type of hazard.

2. Study each type of radiation separately, including methods of protecting personnel and property.

3. Learn the principles necessary for maintaining safety in industries using large quantities of radiation material.

4. Learn how to deal safely with emergency radiation situations.

In summary, our primary purpose in studying radiation is to learn how to live with it so that the great potential benefits can be obtained without unduly jeopardizing personal safety.

4

The Radiation Accident and Fire Record

Fire loss statistics, covering tens of thousands of fires, are maintained by the National Fire Protection Association, the Factory Mutual System, and major government agencies. These records, which have been kept for many years, prove conclusively that it is neglect of fundamentals and not complex special hazards that are responsible for almost all large fires and fire deaths. A careful reading of any fire journal shows that the same causes for starting fires, for converting small fires into large fires, and for fire deaths are repeated again and again. Some common reasons for fire disasters are:

1. Sprinkler control valve shut and forgotten.

2. Fire doors blocked open or inoperative.

3. Open containers or defective safety cans used for flammable liquids.

4. Proper humidity not maintained.

5. Wrong type and/or poor maintenance of electrical equipment.

6. Substandard exits.

The only comparable statistics on fires involving radioactive materials are those gathered and maintained by the United States Atomic Energy Commission (recently replaced by other federal agencies). In the fall of 1971 the AEC issued a publication recording all fires and accidents occurring within its jurisdiction over a twenty-seven-year period. A review of this publication provides ample proof that fires involving radioactive materials do not have to be deadly. It is also clear that for all practical purposes the large fire losses resulted from neglect of fundamentals and not from the presence of radioactive materials. At this point a look at the record is in order.

Fatalities

During the period in question fatalities at operating AEC facilities, at AEC facilities under construction, and in AEC programs at other governmental facilities totaled 295. Of these casualties 292 occurred from such common causes as falling objects, vehicle accidents, electric shock, burns, and asphyxiation. Only three of the deaths could be attributed to radiation. Here are brief reviews of the circumstances surrounding each of these three deaths.

Case 1. An employee, working alone in a laboratory, was stacking blocks of tamper material around a mass of fissionable material that was nearing a critical configuration. He accidentally dropped a tamper block on top of the experimental setup. Observing a "blue glow" the employee proceeded to disassemble the critical material and its tamper. In so doing, he received a heavy radiation dosage from which he died twenty-eight days later. A guard stationed a few feet away suffered no permanent injury. Since there was no danger of a nuclear explosion or of a serious fire, manual efforts to disassemble the experiment should not have been attempted. (There was no fire and there was no explosion.)

Case 2. A scientist was demonstrating the technique of critical assembly. Basically, two hollow hemispheres, one of which contained fissionable material, were slowly being brought together when a screwdriver holding the shells apart slipped. The shells came together and a criticality incident occurred immediately. The scientist performing the experiment received a heavy radiation dosage from which he died a few days later. There was neither a fire nor an explosion.

Case 3. A chemical operator introduced what he believed to be a dilute plutonium solution from one tank to another which was known to contain plutonium in an emulsion. When the employee started the motor driving a propeller type stirrer, a criticality incident occurred. The operator received a massive dose of radiation from which he died a few hours later. A careful check after the incident showed that the plutonium level was about ten times what it should have been at any stage of the operation. Again, there was neither a fire nor an explosion.

Radiation Injuries

Twenty-four years of records covering over 1,600,000 man-years at AEC facilities show that 99.8 percent of employees received an annual radiation dosage of 5 rem or less (for 94.8 percent of the employees the rate was one rem or less; *rem* will be defined later). It is a medically accepted fact that the body's natural defenses will repair radiation damage when the dosage is 5 rem or less per year.

During the period surveyed, the radiation dosage exceeded 15 rem in one year in only 44 instances. These exposures resulted in 38 lost-time incidents as follows: (1)

3 employees died, as cited above, (2) 3 employees required amputation of a finger, (3) 2 received permanent partial disabilities, and (4) the remaining affected employees received minor temporary or no injuries. (It should be noted at this point that a once-in-a-lifetime dosage of 25 rem is considered to be medically acceptable when conditions warrant taking such a risk.) In these instances, also, fire and explosion were absent.

While no overall collection of data covers radiation incidents for all the public and private agencies involved with radiation, the individual records are comparable to those of the AEC. Collectively, these records prove that it is possible to design a control program that provides for virtually complete personnel safety. Nevertheless, because more and more people are coming into contact with operations involving radiation, it is not enough merely to lean on the past outstanding safety record. To prevent carelessness from creeping in, it is essential that the absolute necessity for periodic review of safeguards and the tight enforcement of all rules be repeatedly emphasized.

Property Damage

Unfortunately the personnel safety record has not been duplicated in the case of property. This weakness should not be swept under the rug. It should be examined and appropriate corrective actions taken before a major disaster occurs. The starting point should be a close look at some major fires.

Case 1. Golden, Colorado. Fire of unknown origin (possibly plutonium chips or turnings which are pyrophoric, igniting when exposed to air) started in a storage cabinet at a glove box (a sealed enclosure with a built-in glove for handling radioactive materials). Burning gases swept through a large part of the conveyor system. Damage was estimated at $45 million. A major portion of this loss was the cost of decontaminating the affected area plus plant down time while the work was in process. Typical questions that should be asked for this kind of radiation fire incident include, but are not limited to, the following:

1. Why were combustibles used in the construction of storage cabinets and glove boxes?
2. Why wasn't the conveyor system subdivided by firestops?
3. Could this loss have been prevented by the provision of an installed fire protection system?

Case 2. Paducah, Kentucky. A chemical explosion and fire resulted in a loss of $2,900,000 to building and contents. During the fire approximately 2,000 sprinkler heads operated. Fortunately, no radioactive materials were involved. Had radioactive materials been present, the cost of decontaminating the plant and polluted surrounding areas would have been in the tens of millions of dollars. This incident is a good example of the value of keeping radioactive materials away from hazardous operations.

Case 3. Decatur, Alabama. A three-unit nuclear power plant was provided with first-line and standby safety systems to prevent a reactor melt-down, which could have serious consequences from a radiation standpoint. There was also a room-flooding carbon dioxide (CO_2) system in the cable tray room. The key points to be remembered about this fire are: (1) there was exposed combustible material including some polyurethane; (2) a candle was being used to check the adequacy of the blockage around cables passing from the tray room to the reactor building; and (3) there was a high concentration of cables in the area. With this combination (plus some delay in summoning assistance) the fire, once started, spread rapidly. Damage was heavy and down time was substantial. Appropriate corrective actions have been taken at this facility. However, this fire provided several important lessons for other installations:

1. Eliminate combustible materials in critical areas to the maximum extent possible.

2. Never use an open-flame device to test the tightness of a seal around cables penetrating a wall.

3. Provide appropriate fire protection systems. There is no reason not to use water spray systems over the electrical cables.

4. Because of the critical importance of preventing a core melt-down, make certain that standby safety systems exist and that they cannot all be involved in the same fire. Standby systems that remained functional played an important part in this fire.

The AEC records contain brief summaries of many property fire losses. Of these, most did not involve radioactive materials and the overwhelming majority of fires where radiation was present were controlled with little radiation or property damage. Of the large fires in which radiation played a part, almost all of the portion of the loss caused by radiation was the cost of decontamination. It can also be stated that the reason for the large fires was neglect of fundamentals and not the presence of radioactive materials. In other words, the large fires were due to such factors as presence of combustible materials where only noncombustibles should have been, poor handling of flammable liquids, missing or inadequate fire protection systems, and structural defects.

The superior AEC loss record is duplicated by the Department of Defense in handling nuclear weapons. Here also sound, strictly enforced regulations have produced an outstanding safety record.

To summarize, the record proves that fires in activities

using radioactive materials can be prevented or contained with minimal hazard to life and property provided:

1. The basic facts of radiation are understood.

2. Sound ground rules for the storage, processing, use, and transportation of radioactive materials are established and enforced.

3. Firefighting forces are properly trained in the special procedures needed to fight radiation fires successfully with the minimum possible hazard to people and property.

5

External Radiation

As previously noted, we must deal with two separate and distinct radiation hazard problems—radiation from sources outside the body and radiation from materials taken into the body. Of the two, the external radiation problem is by far the most common. It is also the easier of the two problems to solve.

Although there are three types of radiation, only two (gamma and beta) are important insofar as external exposure is concerned. The third type, alpha radiation, can be stopped by a few sheets of paper or by the skin. Alpha radiation, therefore, is unimportant from an external exposure standpoint, although it can be a serious internal exposure hazard. The two important types of external radiation hazards are (1) long-range, highly penetrating gamma radiation and (2) short-range, less penetrating beta radiation.

Penetration properties of alpha, beta, and gamma radiation are shown in Figure 10.

Gamma radiation is very similar to X-rays and is basi-

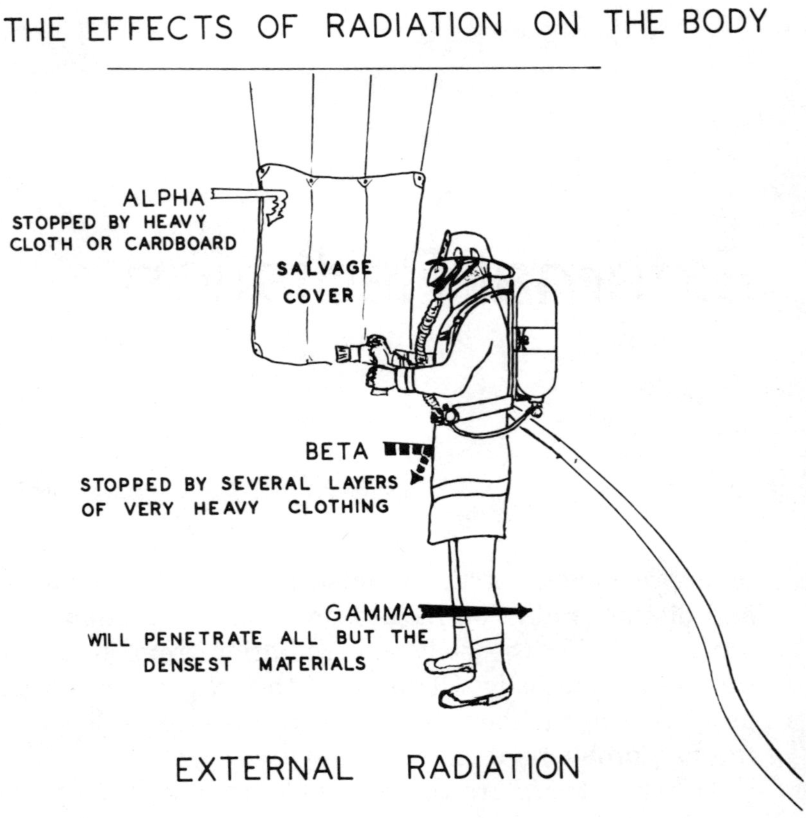

FIGURE 10

cally pure energy. The radioactive source emits this energy in all directions like the rays of the sun. Each energy ray will penetrate the exposed person or other object until its energy is spent. Within the critical range the rays will either strike one or more atoms of living tissue and expend its energy on them (this is where radiation damage occurs) or pass through the body without striking any living tissue and then expend its energy on other objects beyond the body. These latter rays do no harm to the exposed person.

How Radiation Damage Occurs

Observe the two men in Figure 11. The rays coming in their direction from the radiation source penetrate the first man to varying depths and others pass completely through him without touching anything. Of the rays passing through, some will strike targets in the second man at varying depths and others will only pass on

PENETRATING EFFECT OF GAMMA RADIATION
(LONG-RANGE TYPE RADIATION)

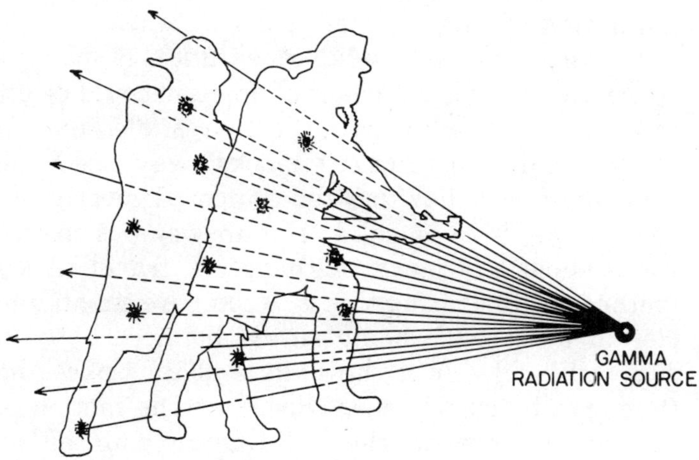

SOME RAYS PASS THROUGH THE BODY WITHOUT CAUSING DAMAGE; OTHERS ARE STOPPED AND THEIR ENERGY SPENT AT VARIOUS DENSITIES OF MATERIALS AND TISSUE.

FIGURE 11

through. These latter rays have done no harm to either man.

It should be noted from Figure 11 that as the object moves away from the source the number of potentially damaging rays to which it is exposed decreases rapidly. This fact is very important in preventing serious radiation damage, as will be demonstrated later in this chapter.

To backtrack for a moment, let's consider how we know that radiation rays can pass through the body without touching anything. X-rays are undoubtedly the best-known proof. If the X-rays, which are similar to gamma rays, were all stopped by the body, there would be no picture for the examining physician. To gain a greater understanding of radiation, let's start with a brief reexamination of what matter really is.

Matter, despite its solid appearance, is mostly empty space. In fact, if the nucleus of an atom could be enlarged to the size of a tennis ball, the nearest electron would be at least eight standard city blocks away (see Figure 12). You might ask how matter composed mostly of empty space appears to be solid. The answer lies in the speed with which the parts of the atom travel. A common method of illustrating this is to compare subatomic particles to white balls being slowly bounced between floor and ceiling. Each of the balls and the spaces between them can be clearly seen. Speed up the motion and the distinction between the balls being bounced and the spaces between them becomes blurred. Speed up the action still more and all distinction disappears; in the place of the bouncing balls the observer sees white pillars. Increase the number and closeness of the objects being bounced, and the viewer will see a white wall instead of pillars.

The next step in seeing how radiation works is to imagine firing a bullet at the rapidly moving objects in what appears to be a solid wall. If only one shot is fired, it

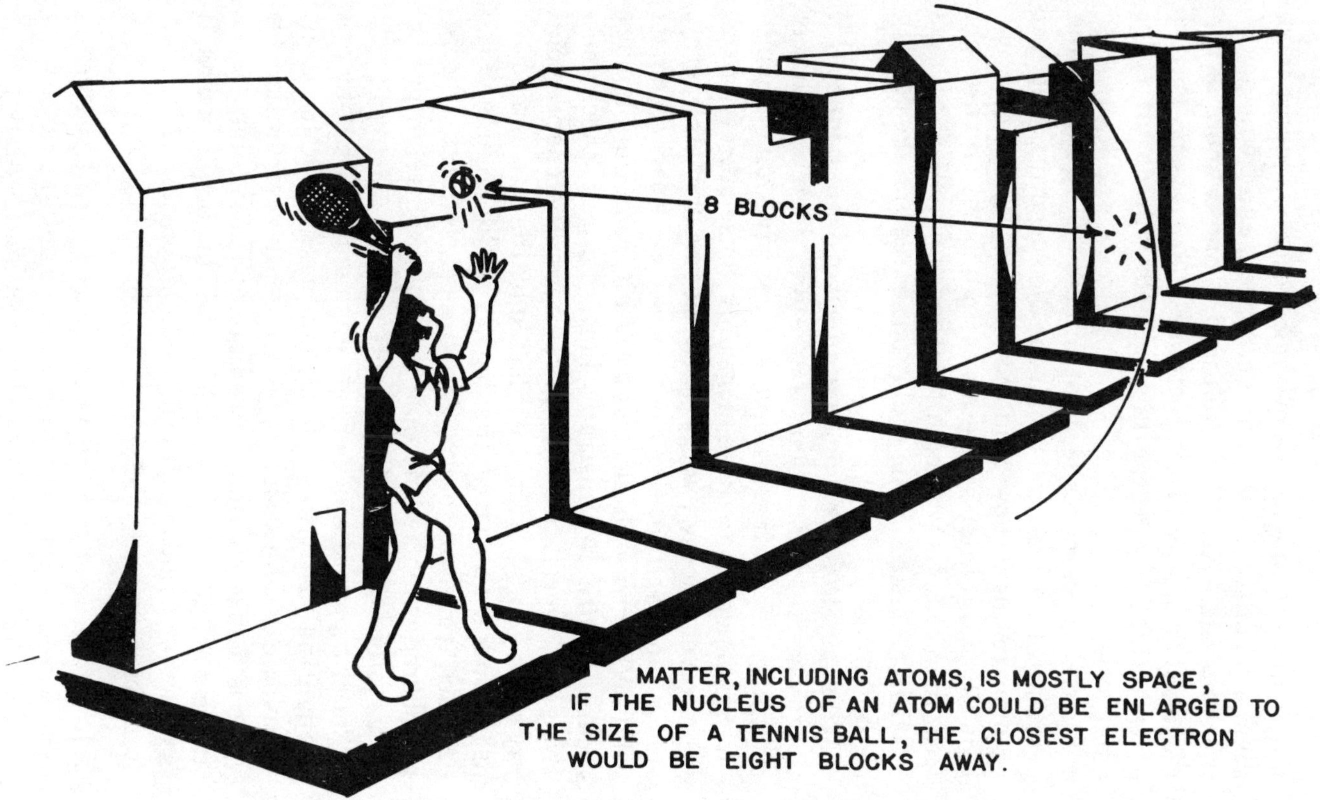

MATTER, INCLUDING ATOMS, IS MOSTLY SPACE, IF THE NUCLEUS OF AN ATOM COULD BE ENLARGED TO THE SIZE OF A TENNIS BALL, THE CLOSEST ELECTRON WOULD BE EIGHT BLOCKS AWAY.

FIGURE 12

might hit any one of the moving objects or might miss them altogether. As the number, speed, and frequency of the shots increases, the likelihood of one or more of the moving objects being hit increases. There is no way of determining, however, which objects will be hit or in what order.

If a radiation source is substituted for the gun, gamma rays for the bullets, and atoms for the speeding objects, it becomes evident how radiation works. Each ray individually will either strike an atom or pass on through. How many atoms will be hit depends on how many rays are being emitted by the radiation source and the density of the atoms.

When an atom is hit by a gamma ray, what happens? If the ray strikes the nucleus of an atom, the probability of anything happening is remote. Few rays have enough energy to penetrate these incredibly dense bits of matter. If, however, the ray, while it still has most of its energy, strikes an electron, the electron will be knocked out of orbit. When enough electrons are hit, radiation damage occurs. Figure 13 illustrates how rays can knock electrons out of orbit. As the figure shows, the chances of a ray hitting an electron are affected by the size and power of the radiation source, the density of the atom, and the distance from the source. All these factors must be remembered when setting up a radiation control or protection system (time, distance, shielding).

For the purposes of persons engaged in fire safety or firefighting activities, it is not necessary to study radiological tables showing dose depth distribution for various degrees of gamma radiation. These tables are of interest primarily to medical specialists and physicists. It is important, however, for fire safety personnel to understand certain basic facts:

1. Radiation exposure is comprised of all sources of background radiation with which mankind has always lived plus radiation introduced by man's activities.

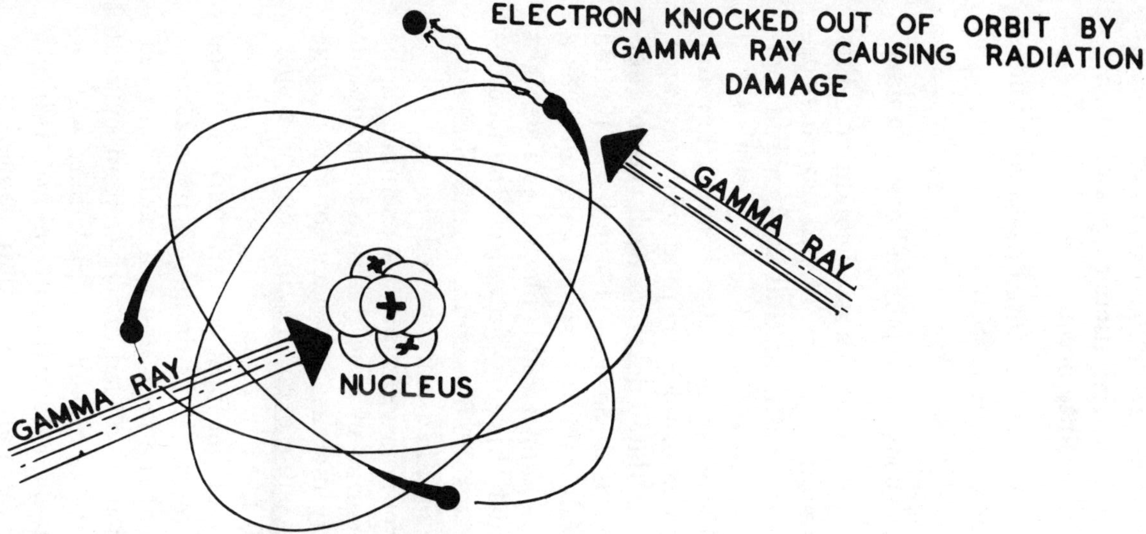

MOST RAYS WILL NOT PENETRATE THE DENSITY OF THE NUCLEUS — THE PROBABILITY OF DAMAGE IS REMOTE.

EFFECTS OF RADIATION ON ATOMS

FIGURE 13

2. While background radiation usually is only incidental, levels of background radiation vary and in some areas it must be taken into account.

3. Manmade radiation can be more or less continuous at a uniform rate, intermittent at a steady rate, or more or less continuous at a variable rate.

4. Within limits the body can repair radiation damage.

While not exact replicas of operating conditions, Figure 14 provides a simplified picture of two typical situations. In each case it was assumed that background radiation was sufficiently above normal to be included.

In part *A* it has been assumed that both background and manmade radiation are occurring at uniform rates. Curves showing accumulated doses for each type of radiation and the combined dosages have been plotted against time. The body radiation repair capability for a fixed period of time is known and is plotted as a straight line. The point at which this curve intersects the accumulated dosage curve fixes the maximum permissible exposure time during a specified work period.

In part *B* background radiation was assumed to be uniform, but for a period of time the rate of manmade radiation was greatly increased. Because of this change the maximum permissible exposure time during a specified work period is greatly reduced. The area labeled "critical time" is the one of most concern and requiring the most careful judgment. If the damage is not too great, if the work is highly beneficial, and if a radiation exposure rate in the ensuing period is nonexistent or well below body capability to repair, the risk may well be acceptable. When the damage in the critical area is likely to be massive and normal body repair capability inadequate, the risk could be justified in only a very few instances (national emergency multiple life rescue, for example). There is no simple, single answer to the ques-

FIGURE 14

EXAMPLES OF RADIATION EXPOSURE VS. BODY REPAIR CAPABILITY

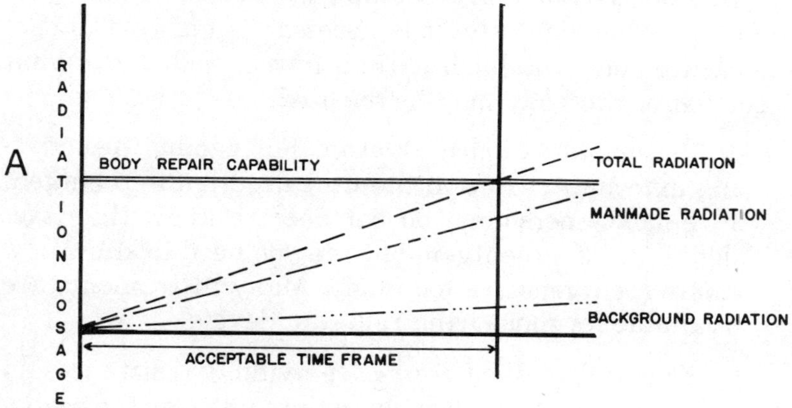

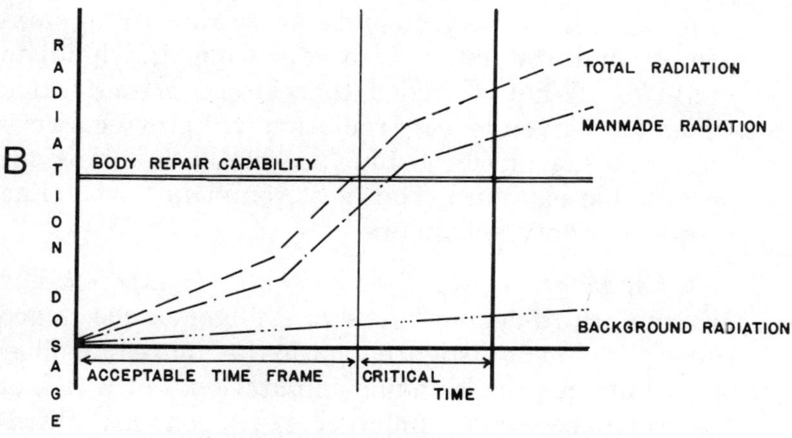

USE THE SAME RADIATION RATES EXCEPT FOR THE PERIOD WHERE RATE WAS GREATLY INCREASED.

tion, "How much radiation exposure is acceptable?" Excess radiation exposure may be a one-time incident or several incidents separated by varying time intervals. It is, however, possible to put any hazard into proper perspective by understanding what it is, how to measure it, and what the basic safeguards are.

Establishing Exposure Standards

In order to establish acceptable standards for permissible radiation exposure, it is necessary to have a unit of measurement, establish a time frame, and know what portion of the body may be exposed.

1. *Have a unit of measurement.* For gamma (penetrating external) radiation the unit used is the roentgen. Fire safety personnel do not need to know the exact definition of a roentgen, but they do need to know how many roentgens are too many. Many instruments are available for measuring radiation levels.

2. *Establish a time frame.* Individual resistances to radiation vary considerably in the same way that resistances to disease, heat, cold, electricity, chemicals, injuries, and so on vary. Body recovery rates from damage of whatever nature also vary from individual to individual. What is needed, therefore, is a time frame which for a given level of radiation will provide a very important key in establishing acceptable standards for permissible exposure. The most commonly used time frame is twenty-four hours.

3. *Know what portion of the body may be exposed.* For example, third-degree burns to a finger or toe might cause permanent disfigurement to the injured member or perhaps require a minor amputation, but a loss of life would be almost impossible. By contrast, third-degree burns to a critical area or to more than half the body could well result in permanent incapacitation or death. The effects of radiation are very similar. An exposure of 1,000 roentgens to a finger or toe for a very short period would undoubtedly result in permanent injury and maybe amputation of the exposed digit plus

possible minor injury to immediately adjacent areas. It would not normally cause injury beyond that area. By contrast, a 1,000-roentgen exposure to the whole body delivered in a short period of time would in all probability cause death.

Because of the obvious effect of time in the measurement of the intensity of radiation exposure, it is essential that the phrase "short period of time" be defined. For the purposes of this text a short period of time is defined as twenty-four hours or less. Time intervals are important for any realistic appraisal of the effects of radiation.

Despite large-scale use of radioactive materials for many years, very few fatalities or permanent injuries have occurred. Due to the lack of a statistical base (a tribute to existing safety programs), it is necessary to establish exposure standards on an empirical basis. Empirical standards, based on observations in the absence of an adequate statistical sample, should be and have been conservative.

No one knows the exact amount of radiation needed to kill any specific individual or any small group of individuals. The starting point, therefore, should be an estimate of the median—the point at which whole body exposures to penetrating external radiation would be lethal to 50 percent of exposed persons. Estimates of the amount of radiation required to produce a 50 percent casualty rate vary somewhat. Because of the small statistical base, we will use the most conservative figure—an exposure of 500 roentgens in a twenty-four-hour period.

The next figure to determine is the amount of radiation required to cause the first death. Here the statistical base is somewhat larger, and it is reasonable to assume that a radiation level of 200 to 250 roentgens in a twenty-four-hour period would be required to cause the first death.

The next key point is fairly well established. At 100 roentgens in a twenty-four-hour period, symptoms to be expected include nausea, vomiting, fatigue, and varying degrees of sickness or injury. Symptoms would increase in severity as the amount of exposure increased.

At exposures below 100 roentgens, symptoms will diminish until at about 50 roentgens exposure there would only be slight temporary blood changes. Decreasing the exposure level would reduce symptoms still further, until at a level of about 25 roentgens there would be no detectable effect.

Figure 15 illustrates the key radiation levels, based on total body radiation exposure in a twenty-four-hour period. Remember that it is almost impossible for any person to receive a high-level exposure unless simple safety precautions have been violated.

Genetic Effects

Before accepting a position involving a possibility of radiation exposure, an individual might well ask, "Safety programs seem to take care of the large single exposure problem, but what about the effect of repeated small doses over a period of years?" In assessing the effects of repeated small radiation doses over a prolonged period of time and determining how much exposure is permissible, two effects must be considered: (1) the effect on the individual over his or her lifetime, and (2) the effect on society through genetic mutations.

Much heat and very little light has been generated regarding the effect on the lifespan from repeated small doses of radiation, due to limited knowledge of what can be done to control these exposures. If common-sense safety measures are taken, small exposures appear to have little if any effect. Large radiation doses can be avoided by simple, rigidly enforced safety precautions.

Because of the widespread misinformation on the subject, it is proper to consider genetic effects first. In this area the following comments are in order:

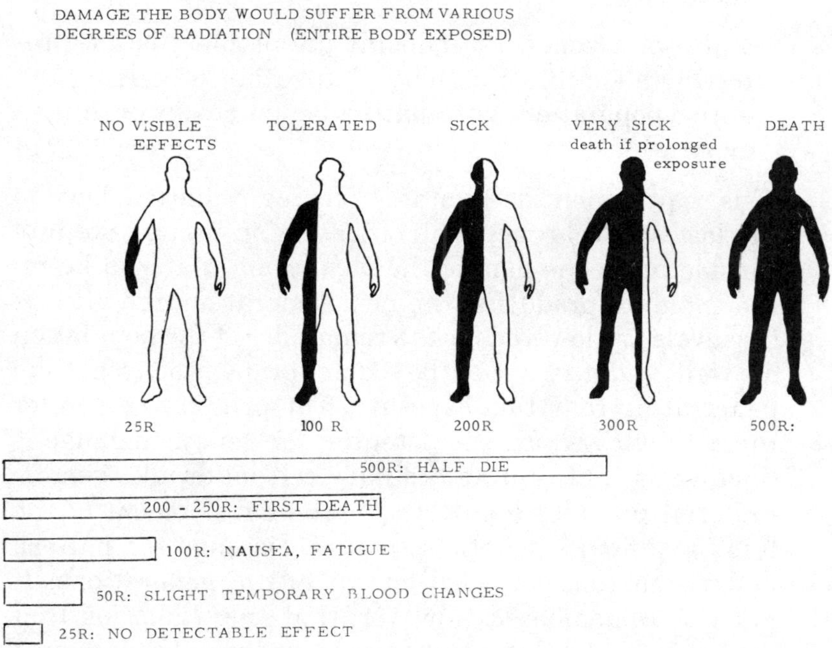

EFFECTS OF EXTERNAL RADIATION
Total Body Exposure within a 24-Hour Period

FIGURE 15

1. Radiation doses in the high range can cause sterility. They do not, however, cause impotence. Repeated small doses do not cause either sterility or impotence.

2. Congenital defects are not necessarily genetic in

origin. About half of such defects are from disease or other causes.

3. Radiation striking genetic material may cause a mutation to appear in the next or some future generation, but nonradiation factors can also cause such changes.

4. From a genetic standpoint the problem is not protecting a specific individual, but rather protecting the entire population, with particular emphasis on future generations.

This requirement necessitates keeping radiation down to the lowest level practicable. If radiation limits take into consideration the part of the population that will be responsible for producing the next generation, the allowable levels are lowered for the remainder of the population as well and there will be no mutants produced from penetrating radiation exposure. The primary reason for the care with which the potential for genetic damage is handled is that, unlike ordinary cell materials, genetic material provides a pattern for the development of the fetus and for its genetic structure. Damage the pattern and it remains damaged for all future generations. It must be remembered, however, that only radiation that strikes reproductive organs can cause genetic damage.

Based on the fact that the body has certain repair capabilities, the population can be divided into four groups in terms of preventing genetic damage:

1. Eighteen or Under. Up to age 18 the body is building its basic framework and is not accumulating appreciable ability to repair radiation damage. Since this group will become a major contributor to future generations, its members should not be exposed to any radiation unless it is medically required.

2. Eighteen to Thirty. Slightly more than half the contributions to future generations are made by members of this group. Being fully grown, they do accumulate body repair capability. Accordingly, these individuals can be exposed to some radiation, provided that over a designated span of time it does not exceed body repair capability.

3. *Thirty to Forty.* Approximately 35 percent of the next generation comes from this group. Permissible radiation exposures will be approximately the same as those for the eighteen-to-thirty group. If members of this group were not exposed to radiation prior to age thirty, the accumulated body repair capability will permit slightly higher exposures.

4. *Over Forty.* Permissible radiation exposures for this group should not vary appreciably from those of younger groups. In emergency situations (life rescue, for example), assuming adequate body repair capability, it is preferable for members of this group to take the emergency actions, as there would be far less likelihood of genetic damage to future generations.

Body Repair Capability

Medical experience has indicated that the body of an adult can repair radiation damage at the rate of 5 rem per year, excluding the years under age eighteen. (The rem is the measure used to indicate the effect of radiation on the body. For penetrating external radiation, one roentgen of energy will have an effect on the body of one rem.) Unused body repair capability is accumulated. It has also been medically noted that an individual can receive dosages of 12 rem per year, provided the acccount is balanced over a reasonable period of time.

Figure 16 is an exposure banking chart based on the foregoing. It shows no dosage received prior to age eighteen and accumulated exposure at the rate of 5 rem per year up to age 70. Figure 16 should not be considered

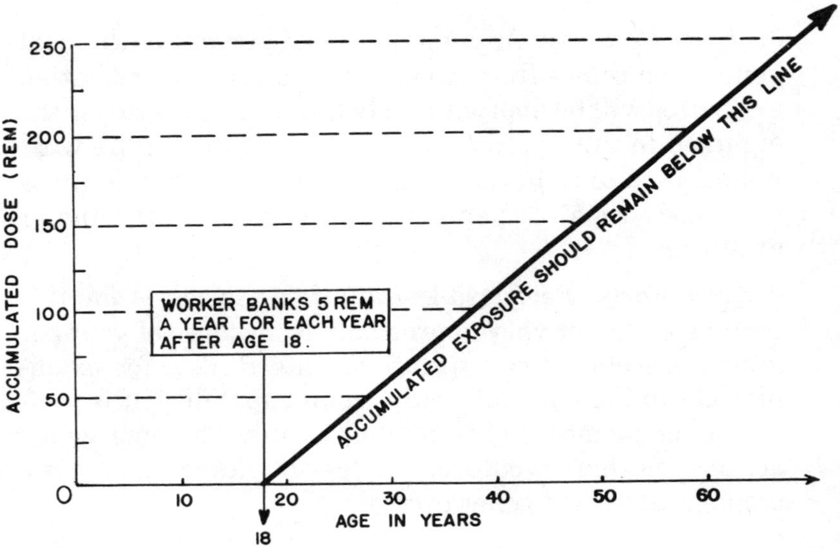

EXPOSURE BANKING CHART (RADIATION WORKERS)

FIGURE 16

representative of actual cases. In practice, radiation exposures are held far below this level.

It is possible to illustrate this radiation banking account, including slight overexposures, by following a new worker from age 22 through 25 (see Table 1). The opening balance of +20 rem reflects the fact that the worker

Table 1

		Balance
4 years @ 5 rem per year		+ 20 rem
Year 1 of work, exposed to 12 rem	−12 rem	
Added during year	+5 rem	
		+13 rem
Year 2 of work, exposed to 11 rem	−11 rem	
Added during year	+5 rem	
		+7 rem
Year 3 of work, exposed to 9 rem	−9 rem	
Added during year	+5 rem	
		+3 rem
Year 4 of work, exposed to 8 rem	−8 rem	
Added during year	+5 rem	0 rem

accumulated the ability to repair radiation damage at the rate of 5 rem per year at ages 18, 19, 20, and 21.

The account has been balanced at the end of four years. Again, in actual practice the dosages received from the amount of radiation exposure permitted are far below these levels, and every effort should be made to keep exposures well below these limits. Occasionally the guidelines for permissible dosage might be slightly exceeded for short periods. This, however, is no more cause for alarm than driving for a few seconds at 58 miles per hour in a 55 mph zone.

Acceptable Exposure Limits

In order to be effective, guidelines must be understood and used properly by management in businesses involving radioactive materials. In most instances it is administratively desirable to break down the work year into quarterly, monthly, and weekly segments. For example, assume a work year of 50 weeks. If the permissible dosage of 5 rem per year is spread uniformly, the rates work out to 5 rem per year, 1.25 rem per quarter, 0.417 rem per month, and 0.1 rem per week. In actual practice, three times a year the weekly rate may be 0.125 rem, and the other 47 weeks the rate may be 0.06 rem. This works out at 3.2 rem per year and a maximum of 0.92 per quarter. The slight overdose is meaningless and is no reason to forgo essential X-rays.

Anytime a radiation dosage over the amount shown in the guideposts occurs, however, an immediate investigation should be undertaken—not because of any hazard to the employee but to make certain that no breakdown in mandatory procedures has occurred or is starting. This is no different from monitoring systems for the control of poisonous gases, liquids, and solids or the control of bacteria in the water supply.

In addition to establishing specific radiation exposure limits, management must establish procedures for keeping current the exposure record of each employee in a radiation area. No individual should be assigned to a radiation exposure area or transferred from one such area to another, unless the record shows it to be safe.

Over and above control systems and record keeping, management has an obligation to keep employees informed of the hazards and the reasons why safety rules must be rigidly followed. This is the final key to having a safe operation.

Protection against Gamma Radiation

Personnel can be protected against penetrating external radiation exposure by three means, used depending on the radiation level involved. The three keys are time, distance, and shielding:

1. Time. Place limits on the length of time a given exposure is acceptable.

2. Distance. Control the distance between personnel and the radiation source.

3. Shielding. Place absorbing material between personnel and the radiation source.

None of the protective measures is ever used alone. Time is always a factor in any program for protection against radiation, and distance, shielding, or both may be required in addition. Let's review each item separately.

Effect of Time on Exposure

Understanding the effect of time on radiation exposure is a simple matter. Assume an individual is standing in a radiation area where the source gives off 2,400 milliroentgens in a 24-hour period. (*Milli-* means a thousandth, so a milliroentgen is one thousandth of a roentgen.) This translates to a body dosage of 2,400 millirem during the same period. In 8 hours the dosage would be 800 millirem, in 4 hours 400 millirem, and in one hour 100 millirem. The effect of time on exposure is shown graphically in Figure 17.

Management uses time as a key factor in planning work schedules and in administratively controlling exposure of individuals. The first step is to have all preplanning of activities occur completely outside the hazard area. This practice minimizes exposure and assures that time spent in the hazard area will be productive. The second step is to establish safety rules assuring that no individual is unduly exposed to radiation hazards. Assume that all other possible precautions have been taken and the remaining gamma radiation penetrating the work area is 250 milliroentgens per workday. To simplify the problem, assume that the radiation energy output is constant and that it has been determined that at no time will any individual be placed in a situation where the dosage received will exceed 100 millirem (the effect of 100 milliroentgens' exposure) per week. Exposure to 250 milliroentgens per 8-hour workday is the equivalent of

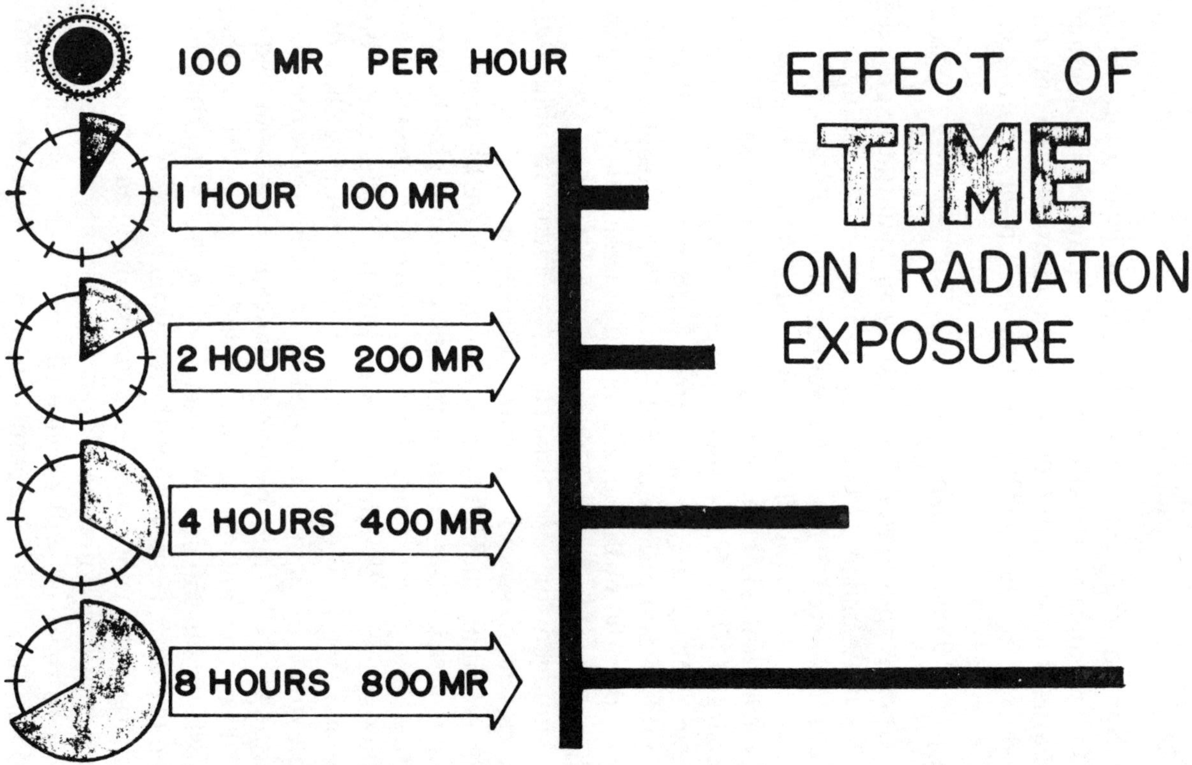

100 MR PER HOUR
EFFECT OF
TIME
ON RADIATION
EXPOSURE
1 HOUR 100 MR
2 HOURS 200 MR
4 HOURS 400 MR
8 HOURS 800 MR

FIGURE 17

100 milliroentgens in 3 hours, 12 minutes. For practical purposes the permissible time per week in the hazard areas is 3 hours. Available time could be allocated in any number of ways, some of which are indicated below.

Hours Permitted in Hazard Area

	Monday	**Tuesday**	**Wednesday**	**Thursday**	**Friday**
Course 1	3	0	0	0	0
Course 2	1	1	1	0	0
Course 3	1	½	½	1	0
Course 4	½	½	0	½	1½

It is also possible to use time to insure safety in accidents involving radioactive materials. Assume a highway accident in which a container of radioactive material breaks and gamma radiation at the rate of 4,000 milliroentgens per hour is emitted. Recorded exposure is at the opposite side of the highway from the accident. Should traffic be stopped or kept moving?

At 45 miles per hour a car travels at the rate of 66 feet per second. In 10 seconds the car has traveled 660 feet and the occupants are well out of the range of harmful radiation. Even if the full dosage was received for this 10-second period, the dosage would only be 11.2 millirem, which is an inconsequential dose. If traffic was stopped at or near the scene of the accident, the results would be quite different, and distance would become a factor. If traffic is stopped, it should be kept at a safe distance from the accident.

The accident cited above could also create problems for occupants of nearby buildings. Usually, as will be demonstrated later, safety requirements for building occupants require that consideration be given to distance and shielding as well as time.

Effect of Distance on Exposure

Distance, properly understood and used, is one of the most effective means of minimizing radiation exposure. As an exposed person moves away from the radiation source, the intensity of radiation exposure falls off in accordance with the inverse square law. This means simply that radiation intensity varies inversely with the square of the distance from the source. Figure 18, par-

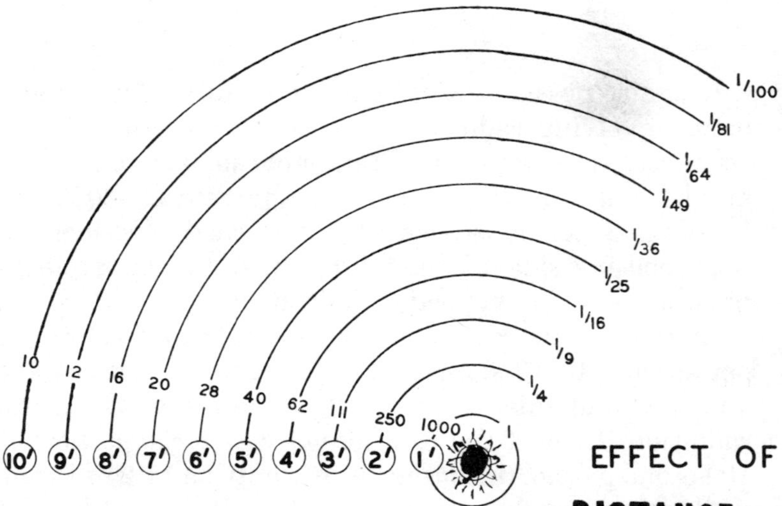

EFFECT OF **DISTANCE** ON RADIATION EXPOSURE (1,000 REM SOURCE)

FIGURE 18-A

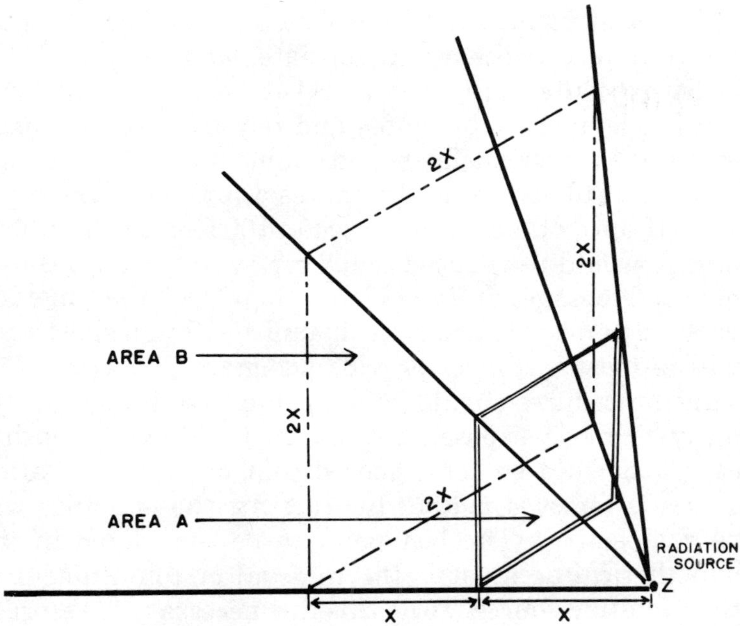

FIGURE 18-B
NOTE: AREA B IS FOUR TIMES THE SIZE OF AREA A

ticularly part *B*, illustrates how the inverse square law works.

In the theoretical accident mentioned earlier, assume that on the side of the highway where the accident occurred traffic is blocked. Under such conditions traffic cannot proceed and time is a factor only in determining the speed with which exposed people must be moved. Here, distance is the means of providing protection against radiation exposure.

Assume that the previously cited 4,000 milliroentgen-per-hour exposure is present at two feet from the source. At four feet from the source the exposure would be 1,000 mr per hour; at 6 feet, 444 mr per hour; at 20 feet, 40 mr; and at 40 feet, 10 mr. If traffic is stopped for 30 minutes the respective dosages would be 2,000 millirem at 2 feet, 222 mr at 6 feet, 20 mr at 20 feet, and 5 mr at 40 feet. (Note: as appropriate, *mr* is used to designate exposure—milliroentgens—or dosage—millirem.) These are appreciable doses and require careful consideration. An hour's delay would double the dosage; cutting the evacuation time to 15 minutes would halve the exposure. If the distance is increased to 100 feet, the half-hour dosage would be reduced to 0.8 mr, which is an inconsequential dosage. It is evident, therefore, that moving exposed persons a moderate distance within a short time will in most instances provide adequate protection. The same procedures should be used in determining how far away to move exposed persons in buildings. In either case, it should be remembered that once the radiation source is removed neither the cars nor the buildings will be radioactive. (The last point to be considered in the hypothetical accident is the removal or containment of the radiation source. To do this it is necessary to resort to the third tool, shielding, which will be discussed later.)

It should be obvious now that distance can be very important in planning industrial work. Since the intensity of radiation decreases rapidly with distance, work procedures should provide for maintaining the maximum possible distance between worker and radiation source that will permit efficient operation. If distance is increased, the time that a worker can safely spend in the hazard area also will be increased.

In industry other than energy production, radiation sources are normally small and provide a starting point for measuring distances and using the inverse square law. For large radiation sources, such as a nuclear reactor, the inverse square law does not start at the reactor. Referring to part *B* of Figure 18, assume the large radiation source to be 4 by 4 feet in area. Putting this area in perspective, the point at which to start measuring distances and using the inverse square law is point *z*. By reducing the size of the source to a point, it becomes possible to determine safe distances from a large radiation source. Failing to do so introduces a serious blunder that can lead to deaths and serious injuries.

Effect of Shielding on Exposure

For many types of activities involving radiation, time and distance are the only tools needed for safe and efficient operations. For other activities with large radiation sources, time and distance are not enough to assure safety. For still other activities, time and distance provide safety but not efficiency. Then the third tool, shielding, comes into play.

The principle of shielding is simply providing a barrier between the body and advancing weapons. In medieval Europe iron shields were used to absorb or blunt the thrust of spears, swords, and arrows. In like manner radiation shields are used to absorb or blunt the energy emitted from the radiation source before the rays can damage the body.

As noted earlier, radiation damage is caused by rays knocking electrons out of orbit. In so doing the rays lose their energy. Providing a material with a large number of electrons in each atom between the worker and the radiation source sets up the protective barrier. The denser the material is, the more electrons it contains, and the greater the degree of safety for a given thickness of shielding. The greater number of electrons in dense materials compared with light materials provides far more targets, and therefore far more rays will be stopped. Lead is a more effective shield than iron; iron is more effective than concrete; concrete more than wood; and wood more than water. Figure 19 illustrates relative shielding efficiencies.

Since matter is mostly empty space, it is impossible to plan an absolute barrier to all radiation. Some rays will penetrate any practical barrier without hitting anything.

RELATIVE EFFICIENCY OF VARIOUS SHIELDING MATERIALS

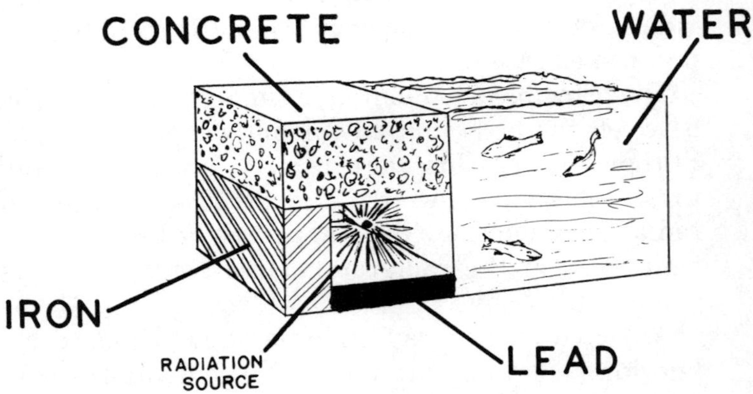

FIGURE 19

Shielding, therefore, is designed to stop most of the radiation so that the amount that does get through is harmless in terms of possible damage.

The unit of measurement used in calculating shielding is the half-value layer—in other words, the amount of material required to stop half of the radiation. Adding a second layer of the same thickness will stop half of the radiation that has penetrated the preceding layer. Again, matter is mostly empty space and there is no way to stop radiation entirely. The purpose of the successive layers of shielding is to reduce the amount of radiation exposure until it is within permissible limits. The effect of adding successive half-value layers of shielding material is shown in Figure 20.

EFFECTS HALF-VALUE LAYERS
OF SHIELDING

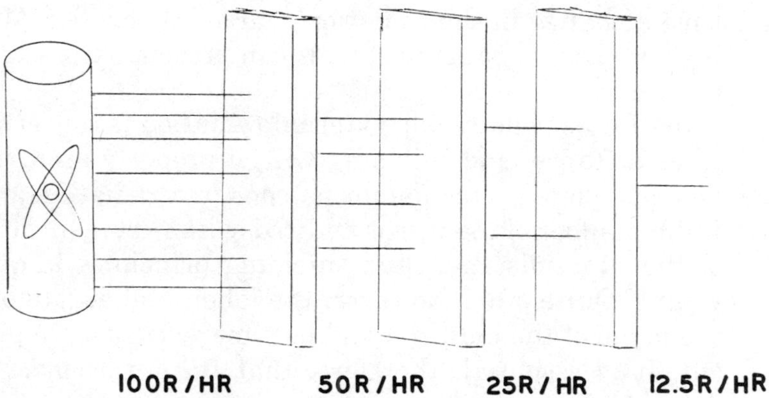

IF THE HALF—VALUE LAYER OF A MATERIAL IS ONE INCH, A ONE-INCH-THICK SHIELD WILL STOP 50% OF RADIATION, A TWO-INCH SHIELD, 75%, A THREE-INCH SHIELD 87.5%; AND SO ON.

FIGURE 20

Selection of shielding must be based on knowledge of varying types of radiation and varying densities of materials. For example, assume a specific intensity of radiation and a requirement to use concrete as a shield. There are many types of concrete with substantial differences in their densities. It is conceivable that a piece of one kind of concrete one foot thick might be entirely adequate, while another type might need to be sixteen inches thick to provide the necessary protection.

Shielding for industrial, utility, and medical exposures is normally designed by specially trained radiological technicians. Personnel in the fire safety field are rarely involved in shielding design; they are primarily concerned with assuring that needed shields are present, that they are maintained, and that the need for shield revisions has been reviewed by a qualified specialist any time a change in the radiation source is proposed. Missing, cracked, or sagging shields can present serious dangers.

Insofar as penetrating external radiation is concerned, time, distance, and shielding will, if properly used, provide adequate protection for persons in a radiation area. If the number of roentgens emitted is known, application of the time, distance, and shielding principles is quite simple. Quite often, however, the label might state only the name of the material and how many curies are present. It is essential, therefore, that the term *curies* be defined.

The amount of a radioactive material that is present is expressed in *curies, millicuries* (thousandths of curies), or microcuries (millionths of curies). Curies are a measure of the rate at which atoms are disintegrating. For this reason one curie of a highly radioactive material might weigh only a fraction of an ounce, while one curie of a slightly radioactive material could weigh tons.

Curies do not indicate radiation hazards or define the type of radiation being emitted; they only tell the rate of atomic disintegration for a particular piece of material. To determine the radiation hazard it is necessary to know how many curies are present and what the material is. The importance of this knowledge is evident from the following comparison. One atom of cobalt 60 on disintegrating releases two gamma rays; one atom of radioactive iron gives off one gamma ray; and one atom of phosphorus 32 gives off no gamma rays (radioactive phosphorus gives off beta rays). Therefore, when roentgens are being measured, one curie of cobalt 60 will emit twice as many roentgens as one curie of radioactive iron, and one curie of phosphorus 32 will emit no roentgens. Table C of Appendix B lists comparative gamma radiation levels for some common radioisotopes. Tables A and B of Appendix B indicate types of radiation emitted by common radioisotopes and heavy elements.

For any given situation, time, distance, shielding, and curie information can be put together to set up a safe operation as illustrated by the following problems. Since cobalt 60 is probably the most widely used radioisotope, it will be used in each of the problems.

One curie of cobalt 60 emits 14.31 roentgens per hour at one foot from the source. At five feet from the source the rate is 14.31 ÷ 25 or 0.57 roentgens per hour; at 10 feet it is 0.143 roentgens per hour; and at 15 feet it is 63 milliroentgens per hour.

Problem No. 1

A worker is located 5 feet from a 5-curie source of cobalt 60 and remains there for one hour. What is his exposure?

5 curies cobalt 60 $14.31 \times 5 = 71.55$r per hour at 1 foot
71.55r per hour = $(71.55 \div 25)$ rem per hour at 5 feet

Worker is exposed to radiation of 2.86 roentgens per hour. Without shielding the worker would receive a dosage of 2.86 rem per hour.

Problem 2

A worker is located 5 feet from a 10-curie source of cobalt 60, remaining there for two hours. How many half-value layers of shielding are required to bring the potential dosage per hour to below one rem?

10 curies cobalt 60 10 × 14.31 = 143.1r per hour at 1 foot
143.1r per hour = 143.1 ÷ 25 or 5.72r per hour at 5 feet.

In two hours radiation would be 11.44 roentgens or a potential worker dose of 11.44 rem

After first half-value shielding layer: 5.72 rem
After second half-value shielding layer: 2.86 rem
After third half-value shielding layer: 1.43 rem
After fourth half-value shielding layer: 0.76 rem

Four half-value layers of shielding are required to bring the potential dosage to below one rem.

In addition to gamma radiation, there is a second type of external radiation called beta radiation, which has far less penetrating power and a much shorter range. Materials emitting only beta rays are an external radiation hazard only when individuals come extremely close to them or handle them and then fail to wash off deposits that may be in contact with the body. Clothing should also be decontaminated as necessary.

Some radioactive materials can be handled with the bare hands. Other radioactive materials can be closely approached but not handled. Still others are extremely hazardous. It is impossible to tell from looking at a piece of radioactive material what kind of rays it is emitting or how safe it is. Until one definitely knows what the material is, the kind of rays being emitted, and the radiation intensity, the wisest course is to assume a major radiation hazard. Such materials should not be closely approached or handled until there is proof that it is safe to do so.

Defining Radiation Terms

The terms used for external and internal radiation must be understood if the various types of radiation are to be handled properly. Some of these terms have already been used, but for comparative purposes we will define them again in more detail.

Alpha Radiation. Alpha particles consist of slow-moving, heavy particles, each containing two protons and two neutrons (identical with the nucleus of a helium atom) and having a positive charge. Alpha radiation loses its energy in 3 inches of air and cannot penetrate the outer layer of skin. As an external radiation force alpha particles are nonpenetrating and nonhazardous. If ingested or inhaled, they become a dangerous internal health hazard.

Beta Radiation. Beta particles are high-speed electrons with a negative charge. Beta radiation, depending on the source, is capable of passing through 10 to 100 feet of air. Few beta particles have sufficient energy to penetrate more than one-half inch of body tissue. Unusually strong beta radiation may be able to penetrate up to one inch of aluminum. Full body exposure to a large beta source is a health hazard.

Gamma Radiation. Gamma radiation is pure energy. It is similar to X-rays and can penetrate several hundred feet of air and up to 3 feet of dense concrete.

Roentgen. The roentgen is a unit used to measure only X-ray or gamma radiation in air.

Rad. The rad measures the absorbed dose of any type of radiation. A rad of one type of radiation may have substantially more effect on the body than a rad of another type of radiation.

Relative Biological Effectiveness (RBE). This term measures the relative biological effectiveness of the various types of radiation. It is particularly important to be aware of this when considering internal radiation exposures.

Rem. The rem is a unit of radiation dose, making it possible to express radiation exposures in one standard term. Table 2 illustrates how these terms are used.

As you have seen, external radiation hazards are easy to control. We have accurate means for measuring the hazard and the amount of exposure that has taken place, and positive means are available to control exposures. If, as in the case of an accident, no measuring instruments are immediately available, it is easy to reconstruct the event. Effects of accidental exposure also show up quickly, and corrective actions can be taken.

Before moving on to internal radiation hazards, some brief comments on one commonly held misconception are in order. Many, perhaps most, people believe that individuals who work with radioactive materials or who have been exposed to industrial or accidental radiation will become radioactive. Strangely enough, this reasoning is not applied to X-rays, although X-rays and gamma radiation are similar. This misconception needs to be corrected. Materials are not made radioactive by exposure to radiation. For a material to become radioactive, it is necessary for the atom nucleus (where radioactivity

Table 2

Type of Radiation	**Rad Dosage**		**RBE**		**Rem Dosage**
Gamma	1	×	1	=	1
Beta (1 me v)	1	×	1*	=	1
Alpha (1 me v)	1	×	20*	=	20

* Variable, depends on material. Note that alpha radiation, although not an external radiation hazard, can do twenty times as much biological damage as the same number of rads of gamma radiation.

starts) to be penetrated. Alpha, beta, and gamma radiation cannot significantly penetrate the nucleus.

Some radioactive materials emit only one type of radiation; some emit two types of radiation simultaneously; some emit all three types at the same time; and still others switch from one type of radiation to another as atom disintegration progresses. These various possibilities make it essential that involved personnel know what material they are dealing with. In knowledge there is safety.

6

Internal Radiation

While the external radiation problem is relatively uncomplicated, internal radiation problems include many factors that are not present when the radiation source is outside the body. One item of difference, in particular, is critically important. An individual can move away from an external radiation source, place a *barrier* or shield between himself and the radiation, or limit the time of exposure. None of these safeguards is available when the radiation source is inside the body.

Radioactive Materials in the Body

Probably the most practical approach to the internal radiation problem for fire safety personnel is to ask and answer two basic questions, then review available means for preventing radiation sources from entering the body. The two basic questions are:

1. How do radioactive materials enter the body? There are four means: breathing, swallowing, skin breaks, and skin absorption.

2. How long will radioactive materials remain in the body? Each case must be individually considered, starting with a basic understanding of what can happen when the entry occurs by breathing, by swallowing, through skin breaks, or by skin absorption.

Breathing or Inhalation. While most material inhaled is almost immediately exhaled, some will remain. This residue may be deposited in the lungs, lodge in the nostrils, or be dissolved or absorbed in the saliva for eventual entry into the digestive system. It should, therefore, be apparent that it is vitally important to know (1) what type of radioactive material has been deposited, (2) what type of radiation is being emitted, and (3) the biological half-life of the material (explained later in this chapter).

Swallowing or Ingestion. Materials that are ingested fall into two classes: (1) those that will be absorbed in the digestive system, and (2) those that will not be absorbed in the digestive system. Materials that are not absorbed will normally pass through the body quite rapidly and be expelled by standard digestive processes.

Materials that are absorbed in the digestive system pass into the bloodstream and are carried throughout the body. Every organ or part of the body removes from the bloodstream any element or elements that are chemically needed. The sensing and pickup operations are limited to chemical properties; no distinction is made between radioactive and nonradioactive forms of the same element.

As an example of sensing and pickup, bones require calcium for building and repair. Chemically, calcium is very similar to radium. Accordingly, bones that require calcium will pick up nonradioactive calcium, radioactive calcium, or radium from the blood. On the other hand, if calcium is not needed, the unneeded chemicals will be returned to the digestive system and expelled from the body. If a radiation source in the digestive system is large enough to do harm while in the system, there are means for speeding up the ejection process.

Potassium is an excellent example of an element used widely throughout the body. Again, it must be remembered that any part of the body needing potassium makes no distinction between radioactive and nonradioactive potassium

Many elements tend to concentrate in one or two organs. Perhaps the best known example is iodine, which concentrates in the thyroid gland. It is this characteristic that has made many radioactive isotopes so valuable in medicine. Knowing where a radioactive material will lodge is a major step toward the solution of many medical problems.

Some organs or parts of them can be safely removed from the body, while others cannot. Knowing what the material is, where it is, and how much and what kind of radiation is being emitted plays a major role in determining whether surgery or some other medical procedure should be used to eliminate the radiation hazard. Removal of internal radiation is a medical problem. Preventing internal radiation exposure from occurring as the result of fire is, however, the responsibility of fire safety personnel.

Skin Breaks. Radioactive materials can enter the body through cuts or sores, and they can also enter wherever the body is not protected by the outer skin layer. Such material may lodge at or near the point of entry. More commonly, however, the radioactive material will be picked up by the bloodstream and be carried through the body in essentially the same manner as materials picked up through ingestion.

Skin Absorption. Radioactive materials acquired through skin absorption usually remain near the point of entry to the body. This type of entry is the least common.

No matter how the radioactive material gets into the body, it should be obvious that some of the material may:

1. Be expelled almost immediately as by exhaling.

2. Remain in the body for fairly short periods (a few minutes to a few hours).

3. Lodge more or less permanently in a nonvital spot (an organ or part that can be surgically removed).

4. Lodge more or less permanently where surgical removal is not possible.

Radiological and Biological Half-Lives

Taking all these things into consideration, it becomes necessary to know not only the material and the type of radiation, but also the radiological and biological half-lives.

Radiological half-life is the time necessary for the material to lose half its radioactivity. The half-life periods, depending on the material, can vary from minutes to thousands of years. In Table 3 radiation decay for potassium 42 (half-life 12.4 hours) and carbon 14 (half-life 5,800 years) is illustrated.

Note that at 50 hours the potassium radiation rate has been reduced to one-sixteenth of the original while the carbon radiation rate is virtually unchanged. The table shows that from a radiological standpoint a radiation source that might be small enough to leave alone while it

Table 3

	Radiation rate	
Hours	**Potassium 42**	**Carbon 14**
0	100	100
12.4	50	100–
24.8	25	100–
37.2	12.5	100–
49.6	6.25	100–

disintegrates naturally in potassium 42 could be potentially lethal and require surgical removal in carbon 14. Figure 21 illustrates the decay of a radioactive material by half-lives.

DECAY OF RADIOACTIVE MATERIAL WITH A 24-HR. HALF-LIFE

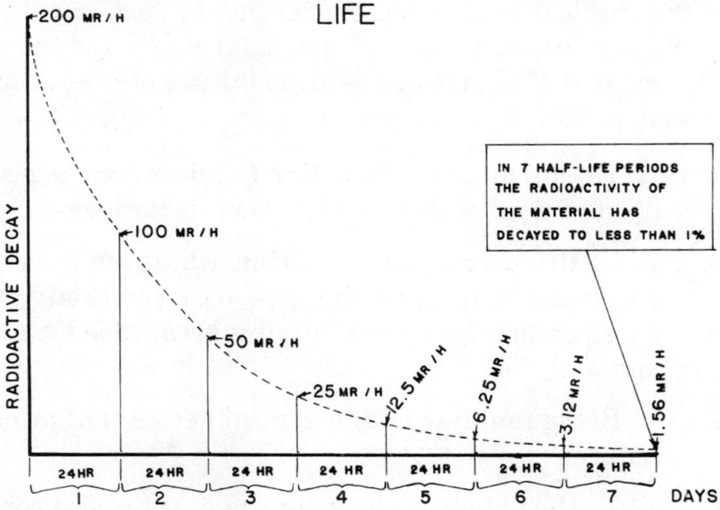

FIGURE 21

Biological half-life is the time required for half the material to be excreted from the body. Some materials pass through rapidly (materials that are not absorbed in the digestive system, for example). Other materials pass through slowly and still others are lodged more or less permanently.

Effective half-life is the unit arrived at when radiological and biological half-lives are combined.

The Role of Fire Safety Personnel

In a case of internal exposure, the actions taken in behalf of the exposed individual are the responsibility of the medical profession. Prevention of such exposures is the joint responsibility of industrial safety and fire safety personnel. The primary responsibilities of fire safety personnel are:

1. Taking appropriate action to minimize the possibility of fire in or near a radiation hazard zone.

2. In the event of fire, taking whatever actions are necessary to prevent the spread of radioactive materials beyond the immediate fire area. (See Chapters 8 and 17.)

3. Being familiar with essential decontamination procedures.

4. To the extent possible, in the event of fire contamination, advising medical personnel of the nature of the radioactive materials and how they entered the body.

As with external radiation, the problem of internal radiation can, under normal industrial operating conditions and during fires, be handled so that the potential for radiation damage to individuals is minimal. The records compiled by the principal users of radioactive materials and by fire departments that have had to deal with fires involving radioactive materials are proof of this fact.

Protection against Internal Radiation

Protection against internal radiation exposure begins in the field of industrial medicine, and much has been done. Standards have been set for permissible body burdens for the various types of radioisotopes in use. Working backward from that, the permissible air concentrations of the various radioactive materials were determined. This step is essential, because in industry radioactive materials enter the body almost entirely by breathing and swallowing. The absorption hazard, where it exists, is virtually 100 percent controlled by the use of gloves and glove boxes (these should be noncombustible). There is no excuse in industry for the entry of radioactive materials through wounds or sores. All that is required in this area is enforcement of standard safety techniques.

While most processes involving radioactive materials are designed so that the possibility of radioactive particles becoming airborne is remote, particles do get into the air in some instances. When they do, it is essential to know their extent. Air samplers are used to pull air through a special filter paper, which is then analyzed in a laboratory for the presence of radioactive particles. From this analysis one of two conclusions can be drawn. The level of radiation present is either below permissible limits or at or above permissible limits.

If the monitored level of radiation is far below permissible limits, standard hygienic precautions may well be enough; as the level rises, additional safety measures become necessary. Whenever the monitoring system shows that the permissible radiation limits have been approached, reached, or passed, an immediate investigation should be made to determine why.

Strict design requirements are imposed by various authorities to prevent a hazardous situation from developing. When a hazard arises despite precautions, however, questions such as the following should be promptly asked.

1. Has the procedure been changed in any way since it was first set up? If so, how and why?

2. If changes have been made and are considered necessary, what additional safeguards are needed to bring airborne radiation back within safe limits?

3. Is the ventilation system functioning properly? If not, why not?

4. Have any changes been made in the materials used in the safety system? (For example, substitute air filters may be more porous than the original type and thus let a higher percentage of radioactive particles through.)

5. Has there been any letdown in enforcement of safety regulations? If so, why?

6. Have cracks or weakened joints that would permit leakage appeared in process equipment?

7. Are containers for radioactive equipment of the proper type? Are they used properly and maintained in good condition?

These and similar questions should be asked and answered and corrective action taken whenever airborne radiation levels approach, reach, or pass designated permissible levels.

Because industrial processes involving radioactive materials are, in most instances, going to continue over a period of years, the internal radiation hazard is a chronic one. For this reason permissible levels of airborne radiation are set so low that it is virtually impossible to inhale sufficient amounts of radioactive material in a single incident or over a reasonable period of time to make up a major part of the allowable body burden. Fireground incidents and certain special processes may, however, develop conditions where the amounts of radioactive materials present will considerably exceed normal permissible industrial limits. For these situations special precautions, which will be examined in later chapters, are necessary.

Basically, the answer to the industrial problem is to develop good techniques and sound regulations commensurate with the risk and to maintain these techniques and enforce the regulations. If this is done (and it generally is), there is little risk to employees, and the jobs of fire safety personnel (particularly fire departments) become much easier and less hazardous in event of fire. Fire safety personnel should, therefore, know the degree of potential hazard and the extent to which essential safeguards and regulations are being used and enforced.

Containment Procedures

No hard and fast rule on safeguards and regulations fits all situations. Requirements will vary widely, depending on the nature of the work being done and the type and intensity of the radiation source. Following are some common procedures used to safeguard personnel against radiation hazards.

1. Processing equipment may be totally enclosed in order to prevent the escape of radioactive material into the working atmosphere or area. Enclosures are used whenever the nature of the work is such that without them the radiation levels would be likely at any time to approach or exceed permissible limits.

2. The radiation hazard may be located in a fire-resistant room or building. Protected types of construction are also acceptable. It should be obvious that no serious radiation hazard should ever be located in a type of building that is subject to a total burnout and/or collapse in event of fire.

3. Smoking is normally forbidden in areas where

radioactive materials are handled with either gloved or bare hands, because of the ease with which radioactive particles could be transferred from hand to mouth and thus inside the body.

4. Eating and drinking facilities (even water coolers) are usually banned from hazard areas because of the ease with which radioactive particles could be transferred from the hand to food or drink and from there to the body or to containers which are removed from the premises.

5. Showers may be required for employees after each day's work. In some instances showers are required every time an employee leaves the area. Such restrictions mean:

a. Provision of a change house, possibly with an airlock. In this intermediate area employees would change from work clothes to street clothes.

b. Special work clothes which remain at the plant and are laundered there.

c. Recovery facilities in the shower and laundry systems to recapture stray radioactive materials.

6. Noncombustible furnishings are generally used to the maximum extent possible.

7. Special permanent vacuum cleaning systems can be used for cleaning floors and walls. (The use of brooms and/or cleaning rags and mops is almost always undesirable.)

8. The speed as well as the direction of airflow can be controlled. The lower the speed, the less likelihood there is of picking up radioactive particles.

9. Proper respiratory equipment can be provided for situations in which personnel may be unavoidably exposed to airborne concentrations of material that may contain radioactive matter. Depending on the nature of

the hazard, equipment may range from simple filter masks to totally enclosing, impervious suits with independent air supply. Presence of such equipment should warn firefighters of their specific needs for protection in event of a fire emergency.

10. There are mandatory requirements that, when containers of radioactive materials are moved, the original covers or their equal shall be securely fastened.

11. There is a strictly enforced rule that all safety devices and procedures will be used as prescribed in plant regulations and that no exceptions will be permitted.

Several of the steps that can be taken to provide safety from potential internal radiation exposure are illustrated in Figures 22 through 28.

Emergency Situations

In spite of all precautions, an occasional employee or firefighter might be exposed to an excessive concentration of airborne radioactive material. If this should happen, it is essential to find out quickly what the material is and how much of it is in the body. This is a job for a radiologist, not for fire safety personnel. The chances of such an overdose, as proven by the record, are extremely small, but the possibility should not be overlooked. The reason the chances are slight is rigid adherence to carefully designed operations, superior maintenance, and tight enforcement of regulations.

Although such a situation is unlikely (given proper design), an emergency in which air handling or processing equipment is disrupted may occur. For example, despite all efforts to site and construct a facility properly, it is conceivable that a seismic shock wave could damage equipment and release radioactive material. In such an

INTERNAL RADIATION IS RECEIVED

RADIOACTIVE MATERIALS.

SEVERAL PRECAUTIONS ARE:

IF RESPIRATOR IS PROVIDED,
USE IT; TAKE CARE OF IT.

FIGURE 22

EXHAUST SYSTEMS MUST BE OPERATED AS DESIGNED

GOOD DESIGN & INSTALLATION CAN BE RUINED BY POOR OPERATION OR MAINTENANCE.

FIGURE 23

COVER CONTAINERS OF "HOT" MATERIALS WHEN BEING HANDLED OR MOVED.

FIGURE 24

KEEP YOUR HANDS OUT OF "HOT" MATERIAL.

FIGURE 25

WASH UP BEFORE YOU EAT AND BEFORE YOU GO HOME.

FIGURE 26

BROOMS STIR UP DUST;
USE
VACUUM
CLEANERS.

FIGURE 27

DO NOT EAT IN AN AREA
WHERE RADIOACTIVE MATERIALS
ARE PROCESSED.

EAT IN THE LUNCH ROOM
WHETHER YOU BRING IT OR BUY IT.

FIGURE 28

event there is no way to make an immediate analysis of the hazard, although the intensity of the radiation can be measured. Assuming radiation levels to be within the acceptable range for a known period of time, properly equipped personnel can enter the hazard area for such purposes as life rescue, decontamination, and so on. At a minimum these individuals should be equipped with self-contained breathing apparatuses. In more dangerous situations they should also be provided with special protective clothing. On leaving the hazard area these individuals should immediately shower and change clothes. All clothing and equipment brought into the hazard area should be left there for later decontaminating. Following these procedures will assure that personnel have not picked up radioactive materials in or on their bodies.

In sum, the internal radiation problem is capable of solution, provided appropriate safety regulations have been prescribed, are understood, and are enforced.

7

Neutron Bombardment

At the beginning of the discussions on external and internal radiation, the statement was made that persons exposed to radiation from radioactive materials as ordinarily used would not become radioactive. The word *ordinarily* should probably be underlined for emphasis. In this area fall better than 99.5 percent of all activities involving radioactive materials. There are reportedly a few types of operations where this may not be true, but they are extremely few in number and are conducted only under the tight control of highly trained scientists. It is not anticipated that fire safety personnel will ever run into this problem. How then do materials get to be radioactive?

First, let's examine briefly one source of artificially induced radiation. In several types of machines, charged particles are fired at a target at tremendously high speeds. Unlike natural radiation, in which the charged particles move in all directions, these particles follow a directional path (like a laser beam compared with an ordinary light bulb). Neutrons can be similarly fired. Under these special conditions the target becomes radioactive. Such situations should, however, pose no problem to fire safety personnel. All these machines require large amounts of electrical power and all are provided with outside electrical controls. In the event of an emergency, fire or otherwise, the main switches are pulled immediately. The production of artificial radiation then stops and the only radioactive material present is the target. These targets present no problem, as their radiation is identical with that of comparable naturally radioactive materials.

Referring back to the structure of the atom, it should be noticed that the proton gives alpha particles their positive charge, that the electron (beta) has a negative charge, and that gamma radiation is pure energy. Except under unusual circumstances, none of these sources can make other materials radioactive. The other major part of the atom, the neutron, has the means to make other materials radioactive.

The neutron is electrically neutral and is therefore neither attracted to nor repelled by either electrons (negative charge) or protons (positive charge). Neutrons are found in the nuclei of all atoms (except normal hydrogen, which has none) and are tied to protons by a strong binding energy. It is possible, as will be demonstrated in Chapter 11 on radiation physics, to dislodge neutrons from the nuclei of some elements.

Free neutrons, being neither attracted to nor repelled by charged particles, leave the nuclei at varying rates of speed and travel until they hit the nuclei of other atoms. At this point one of four things of importance to fire safety personnel could happen.

1. The combination of neutron mass and speed would not be sufficient to penetrate the nucleus.

2. The neutron might strike the nucleus of a nonfertile

atom (one that cannot become fissionable) and nothing would happen.

3. The neutron might penetrate the nucleus of a non-fertile atom and be captured. When this happens, the material becomes radioactive. This is the process whereby radioisotopes such as cobalt 60 are made.

4. The neutron might strike the nucleus of a fissionable element (there are only a few) and split it. This action produces a great deal of energy which can be used to produce power or to create atomic explosions (e.g., nuclear weapons).

The only areas where fire safety personnel are at all likely to come in contact with neutron bombardment are nuclear reactors, atomic weapons, and certain special neutron sources. The risk of any such contact is extremely slight, but it is present and deserves mention.

The part neutrons play in the overall radiation picture will be covered in more detail in Chapter 11, "Radiation Physics for the Layman"; Chapter 12, "Nuclear Power Plants"; and Chapter 17, "Radiation Hazards in Firefighting."

8

The Contamination Problem

In studying radiation contamination hazards the logical starting place for fire safety personnel is to outline certain key facts about radioactive materials:

1. The chemical properties of any radioactive element or compound are identical with those of nonradioactive particles of the same element or compound. This means simply that:
   a. Radioactive materials will burn in the same way and at the same rate as their nonradioactive counterparts: Carbon will burn and so will radioactive carbon 14; and cobalt will not burn and neither will radioactive cobalt 60.
   b. If a given material (such as benzene) gives off vapors that mix with air to form explosive mixtures, so does the material's radioactive counterpart.
   c. Radioactive materials will combine chemically in the same manner as the corresponding nonradioactive materials.

2. The physical properties of radioactive and non-radioactive elements other than atomic weights are identical or nearly so.

a. If gaseous, liquid, or solid particles of a material can become airborne, so can the radioactive forms of the same material.

b. If particles of a compound are soluble in water, so are radioactive particles of the same compound.

3. The amount of radioactivity present remains unchanged by industrial processes, fires, explosions, or natural disasters. For example, benzene in its normal state has six atoms of carbon and six atoms of hydrogen. Assume a fire in which benzene along with other materials burns. Among the products of combustion will be carbon dioxide (carbon from the benzene and oxygen from the air). If these carbon atoms happened to be radioactive, radioactive carbon dioxide would be formed as the benzene burned. From a radiological standpoint, all that has been changed is the location of the radiation. In the new location the radiation will continue to decay at the same rate as it did before the fire.

4. The half-life rate of decay for a radioactive element will vary from seconds to thousands of years. It follows, therefore, that procedures taken by industry to prevent the release of radiation will necessarily be much more stringent when the element has a long half-life. It also follows that in preplanning tactics firefighting forces must take into consideration the half-life potential for contamination.

5. Radiation cannot be neutralized. The acid, alkaline, toxic, and other properties of chemicals can be neutralized or diluted to the point where they become harmless. The removal of radiation contamination, however, can be extremely costly, as will be demon-

strated later. Accordingly, industrial processes and firefighting operations must be designed to hold contamination to a minimum and within as restricted an area as possible.

6. In some industries the loss of control of even small amounts of radioactive material can be very serious, even though employees have or can be given 100 percent protection against personal radiation injury. Some examples of this are: (*a*) manufacture or processing of film, (*b*) manufacture of light-sensitive paper, and (*c*) some foods, drugs, and cosmetics. Such industries must be very careful when radioactive materials in even small amounts are used.

7. Radioactive materials, either natural or artificially produced, vary from elements as light as hydrogen to elements as heavy as uranium, which is nearly nineteen times as heavy as water. In addition, the amount of material needed to sustain a given rate of radiation varies tremendously. To illustrate the point, assume that a situation requires one radiation source of one curie of cobalt and a second source of one curie of sodium. From this data the information in Table 4 can be determined.

Table 4

	Amount Required for 1 Curie	**Half-Life**	**Type of Radiation**
Cobalt 60	880 Micrograms	5.24 Years	Gamma and Beta
Sodium 24	0.12 Micrograms	15.1 Hours	Beta and Gamma

	Flammability	**Reactivity with Water**
Cobalt 60	None	None
Sodium 24	Combustible, Class 1	Violent

It should be obvious that, in terms of contamination, flammability, and reactivity, cobalt usually presents far less of a problem than sodium. Because of its weight, the size of the source required, and its chemical characteristics, widespread contamination by cobalt or by elements with similar properties is unlikely, assuming it is used in its normal solid form. If, however, cobalt is used in a form (such as powder) in which it can get spread around, it can present a major problem because of its long half-life. By contrast, sodium radiation sources become harmless in a few days because of the short half-life.

8. Dry radioactive materials are frequently stored in special metallic capsules that have screwed or soldered tops. These materials normally present little contamination hazard under typical fire exposure conditions in a radiation area, provided the caps are tightly in place. It is necessary, however, to take special precautions to make sure no moisture gets in these capsules regardless of the nature of the radioactive material contained. The reason is that nuclear radiation can convert the water into hydrogen and oxygen. This puts the capsule under strong internal pressure and it may rupture even at normal ambient temperatures. Under fire temperatures the likelihood of capsule rupture would be multiplied enormously.

9. Siting and construction play major roles in the ability to limit the spread of radioactive contamination and in simplifying the task of decontamination. Major factors of importance are:

a. Siting. If appreciable quantities of radioactive materials are to be present and are susceptible to becoming airborne or waterborne, buildings should be located well away from other buildings or important operations where contamination could present a personnel hazard or interfere with major operations.

b. Construction. Buildings should be fire resistive or noncombustible. Floors should be smooth, nonporous, watertight, and easy to clean. Walls and roofs should be as free from ledges and other dust collectors as possible. Interior finish, acoustical and ornamental ceilings, and insulation should be noncombustible and nonporous.

c. Building Height. Preferably, buildings handling radioactive materials should be one story in height and without basements, crawl spaces, or any other below-grade areas. Where this is not possible, the radiation hazard should be confined to the first story and below-grade spaces should be arranged to minimize the contamination hazard. Floors immediately above the radiation hazard areas should preferably be watertight.

d. Drainage. Building design should provide special drainage systems to carry off any sprinkler or hose stream runoff to a safe area such as a holding tank. It might also be necessary to provide curbing between the radiation area and other areas.

e. Cutoffs. While not spelled out in the codes, appropriate cutoffs should be provided between the radiation area and adjoining areas. Such cutoffs, assuming an important radiation area, not only should be able to withstand a total burnout of the exposing area without failure but also should prevent damage to essential shielding. A wall can stop a fire while at the same time transmitting enough heat to cause a lead shield to soften and sag.

10. To minimize the potential for radiation contamination from waterborne radioactive materials, processes involving radioactive materials should be arranged with a minimum of combustible construction and occupancy. Where some combustibles cannot be avoided, automatic sprinkler protection should be pro-

vided. To minimize the number of heads that will operate and to hold the amount of water used to a minimum, the author believes that, in general, an on-off system is desirable in heated areas and a preaction system in unheated areas. In a few areas special auxiliary type extinguishing systems may be in order.

11. Ventilating systems, including filters, should be essentially noncombustible and designed to minimize the possibility of airborne radioactive materials being discharged outdoors or into other areas.

12. Basic controls on radioactive materials are imposed by agencies of the federal government, principally the Nuclear Regulatory Commission, the Energy Research and Development Administration, the Department of Transportation, and the Department of Defense.

In some areas state governments have also set up regulatory agencies to supplement federal agencies. Depending on the type of radiological emergency, any of these agencies can be called upon for assistance. If a weapon is involved, the nearest military base should be called in. For other types of incidents, the regional office of the appropriate agency should be contacted. Fire safety personnel should keep an up-to-date listing of offices in their respective areas where emergency assistance can be obtained.

A second point necessary for control is security. All processes involving appreciable amounts of radioactive materials need to be safeguarded around the clock against thieves, arsonists, vandals, and the like.

Third, fire safety personnel should be aware of the local agencies likely to be involved in the event of a fire or potential fire incident involving radioactive materials. These agencies include police, public health authorities, hospitals, and in some incidents the water and sewer department. There should be continuing liaison between these groups as appropriate.

Examples of Contamination Problems

Fire safety personnel must consider all of the major points just discussed whenever the use of radioactive materials in any quantity is proposed. How far it is necessary to go into each of these areas depends on the nature of the hazard, namely: (1) the type of material (solid, liquid, or gas), (2) the flammability of the material, (3) the amount of material, and (4) how and where the material is to be used. Some examples of possible situations are:

Case 1. Small amounts of solid radioactive material, insufficient to be any hazard to health, become airborne and:

a. Lodge near sensitive radiation measuring instruments or laboratory electronic equipment. The laboratory experiments are ruined.

b. Lodge in a computer. The computer output becomes unreliable.

c. Lodge in an average factory or warehouse. No damage is done.

Case 2. A substantial amount of radioactive material is liberated by an industrial accident or fire:

a. In a fire resistive building with smooth concrete floors, roof, and wall. Occupancy is metal-working. Recovery of the radioactive metal and restoration of production within a reasonably short time is entirely feasible.

b. In a brick, joisted building with hollow-joisted ceilings and furred-out walls. Occupancy is woodworking. Recovery of the radioactive material could be very expensive. In an extreme case (and this has happened), it might be cheaper to abandon the building.

Case 3. Alpha emitters in small quantities may be deposited on the walls, floors, or ceilings. These become a problem only when they become airborne and are ingested, inhaled, or absorbed through skin breaks. It is, however, a potential problem and must be dealt with.

Case 4. Beta emitters may be deposited on building or equipment surfaces and present the same type of potential exposure as alpha emitters. In addition, beta emitters can create high radiation levels near the contaminated surface. Such radiation is hazardous to personnel.

Case 5. Gamma emitters may produce a high radiation level throughout the work areas.

To sum up the contamination problem, there is no single "pat" answer. Each case must be analyzed on its own merits in line with the major areas just outlined. Above all, two questions must be asked and answered if proper prevention and protection measures are to be applied. If radioactive materials can escape from the process: (1) where will they go? And (2) what harm will they do?

Preventing Contamination

Contamination can be prevented in many ways, a number of which have already been mentioned. To these the following commonly used equipment and procedures should be added:

1. Hand and foot counters very similar in operation to the static registers used in hospital operating areas and in some plants handling flammable liquids and explosives can pick up the radiation given off by radioactive particles on employees or their clothing.

2. Strippable covers can be used for tabletops, workbenches, walls, and so on. As required by operating conditions, the covers are periodically removed and

replaced. Contaminated covers are disposed of by radiological experts.

3. Where special work clothes are required, they should be marked by identifying insignia. As necessary, this could mean clothing that is worn to a change room where workers switch to street clothes (work clothes are then laundered at the plant) or it could mean clothing that is restricted to a single area. Both plant and fire department personnel should recognize the insignia.

4. Special handling procedures may be required to move radioactive materials and/or contaminated equipment or supplies from one area to another. The nature of the shielding and/or any special packing or transportation will, of course, depend on the individual situation.

5. Assuming there has been some contamination as the result of a fire or accident, there will almost inevitably be a cleanup problem before the area can be restored to use. The handling of cleanup operations depends on a number of factors: (1) the nature of the contamination, (2) the importance of the operation, and (3) the cost of decontamination. For example:

a. If the contaminating agent has a short half-life, the most practical measure might be to let the area stand idle until the radiation reaches a safe level.
b. If the contaminating agent has a long half-life, some means must be found to remove the source of radiation. The selection of means should be left to specialists in this field.
c. Between the two extremes will be radioactive materials with half-lives varying from perhaps 24 hours to 6 months (these limits are arbitrarily selected for illustration purposes). In this middle range the decision whether to program complete or partial decontamination or to wait for natural decay is strictly a

matter of judgment. All the elements should be weighed in making a decision.

So far, the contamination problem has been viewed primarily from an industrial point of view. This does not mean that these subjects do not directly concern fire departments, plant fire brigades, or plant fire safety personnel. These individuals for their own safety and the safety of others should be aware of the various types of radiation hazards they might encounter in the event of fire. Radiation decontamination after a fire is part of the cost of fire. These costs can be extremely high if mistakes are made in firefighting operations because of lack of knowledge.

At any given plant the local fire safety personnel should be thoroughly familiar with their particular problem and ways to handle it. Fire departments should have a general knowledge of the overall problem in the event of fire. Among other things, this means:

1. Knowing whether or not the plant has a suitable set of rules for the hazard and how well the rules are enforced.

2. Knowing what the radioactive material is and what type of radiation is emitted.

3. Knowing the potential for airborne and waterborne contamination and available facilities for confining the hazard.

4. Having prefire arrangements for working together with plant officials on emergency procedures.

5. Having a departmental preplanning program. At the fireground, only existing knowledge can be put into effect.

A few of the major points in handling the contamination problem are shown in Figures 29 through 32.

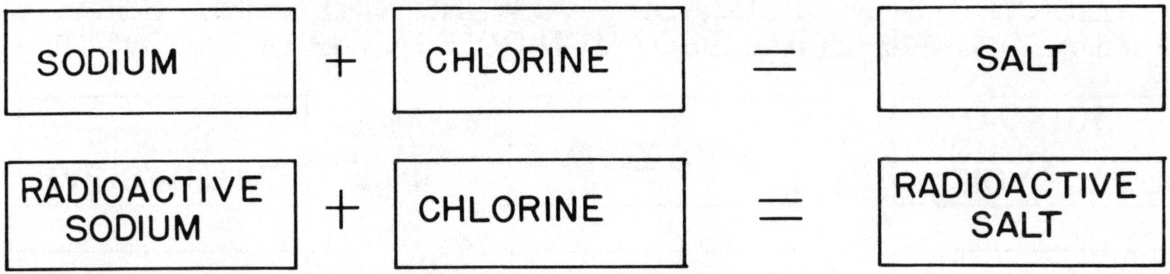

CHEMICAL PROPERTIES DO NOT CHANGE WHEN ELEMENTS BECOME RADIOACTIVE.

FIGURE 29

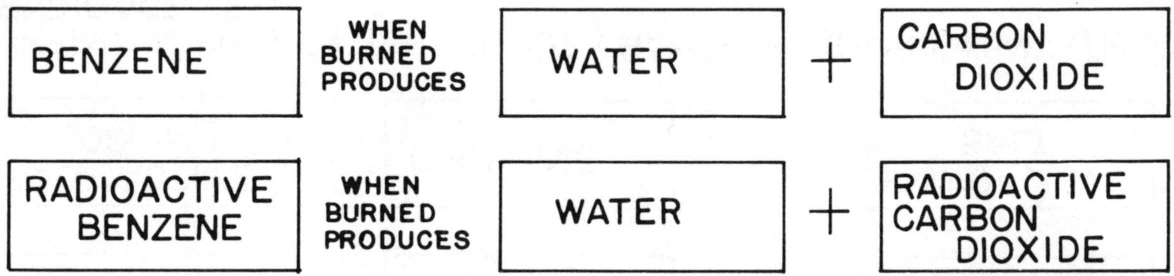

RADIOACTIVE ELEMENTS DO NOT LOSE THEIR RADIOACTIVITY WHEN BURNED. ONLY THE CHEMICAL COMPOSITION IS CHANGED.

FIGURE 30

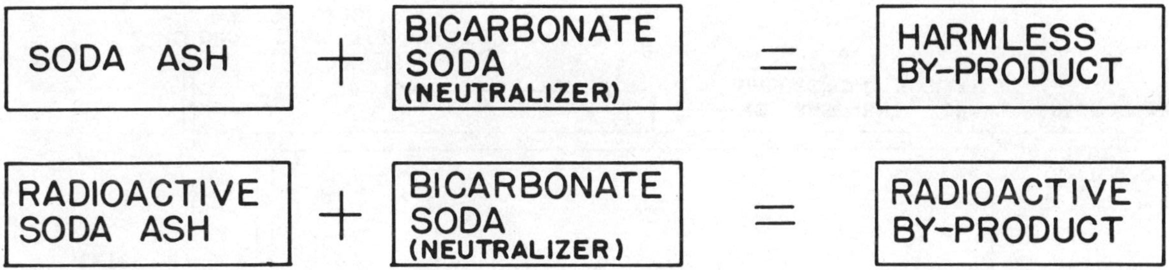

WHEN THE PH FACTOR OF ANY RADIOACTIVE SUBSTANCE IS NEUTRALIZED, THE SUBSTANCE DOES NOT LOSE ITS RADIOACTIVITY.

FIGURE 31

TYPICAL DESIGN FOR A STRUCTURE
HOUSING RADIOACTIVE MATERIALS

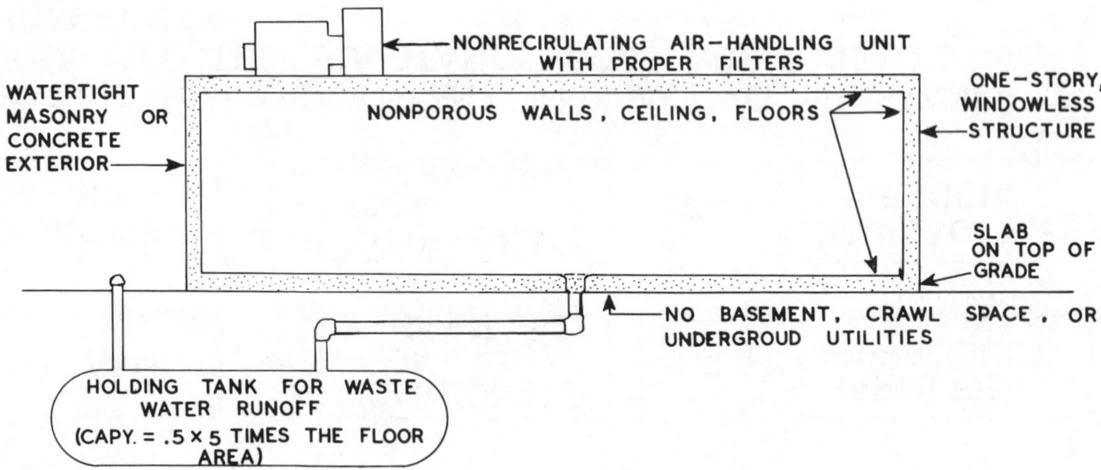

FIGURE 32

In any emergency situation, such as a fire, there are many possibilities for the unintentional spread of radiological contamination to the point where it creates a hazard for people who may have to use the general area and for property far beyond the immediate fire area. These hazards, which are distinct from direct industrial hazards, will be described in Chapter 17, "Radiation Hazards in Firefighting."

9

Transporting Radioactive Materials

Since the beginning of the Atomic Age, millions of packages of radioactive materials have been transported safely. In fact, the volume of commercial shipments has grown to the point that the present rate in the United States alone probably is more than one million packages per year. To date, as far as the author has been able to determine, no deaths or serious injuries have been suffered by the public or common carrier personnel. Damage to property has also been minimal. This outstanding safety record is unquestionably due to the attention given to proper packaging and to the effectiveness of the safety standards required of common carriers.

In addition to shipments of commercial radioactive materials, increasing numbers of nuclear weapons have had to be shipped. Here, too, the transportation of these weapons has an outstanding safety record. For the purposes of this chapter, however, comments will be limited to nonmilitary shipments. Military shipments will be covered in Chapter 14, "Nuclear Weapons."

At present, the great majority of shipments are small to moderate quantities of radioactive materials, many of which have a fairly short half-life. The small size of the packages and in many instances the need to get from producer to user as rapidly as possible inevitably means large-scale use of air freight. In some cases passenger aircraft are also used. Large quantities of radioactive materials and materials with longer half-lives are normally shipped by surface transportation (trains, trucks, and ships).

Regulatory authority for the shipment of radioactive materials in interstate or foreign commerce, except for postal shipments, rests in the Department of Transportation (DOT). Postal shipments are regulated by the U.S. Postal Service. Postal shipments (a small fraction of the total) must conform to essentially the same standards as those set by DOT. Intrastate shipments come under the control of the appropriate state agency by virtue of "agreements" in which the states have adopted uniform regulations for intrastate operations providing for the same packaging, labeling, and marking requirements that are imposed by the Department of Transportation for interstate shipments. It is essential that fire safety personnel be aware of these requirements and know what they mean in terms of fire safety activities.

Numerous federal and state agencies are involved in publishing standards and regulations relative to the transportation and licensing requirements for radioactive materials. However, for the purposes of a fire safety text, the primary concern is the packaging, labeling, and marking requirements of the Department of Transportation.

Shipping Classifications

For shipping purposes, several classifications have been established, the first of which is "Special Form

Radioactive Material." By "special form" is meant materials which, if released from the package:

1. Might present a direct external radiation hazard.

2. Would present little, if any, hazard from radiotoxicity as a result of contamination. This freedom from contamination hazard is due to:
   a. Physical integrity of the material (e.g. a radioactive metal bar).
   b. High integrity in a sealed source.

Figure 33 illustrates three types of special form radioactive material shipments.

"Normal Form Radioactive Materials" include radioactive materials that do not qualify as "special form." Materials in the "normal form" group are further subdivided into seven classes. All seven normal form classes (e.g., waste material, liquids, powders, frangible solids, and so on) and the special form class are further divided by quantity into *A* and *B* groups. *A* and *B* refer to packaging, as explained later. Groupings are given in Table 5.

There is a similar table placing limits for small or exempt quantities, manufactured articles, and devices which are allowed to be shipped through the mail subject to postal regulations. The quantities involved are too small to be of direct concern to fire safety personnel outside of the postal service.

Packaging

Type *A* packaging is designed to prevent the loss of or dispersal of its radioactive contents and to maintain its radiation shield when subjected to normal conditions of transportation. Approved containers have been subject to tests required by DOT to simulate normal conditions. Containers include specification steel drums, wooden boxes, fiberboard boxes, and a general purpose package.

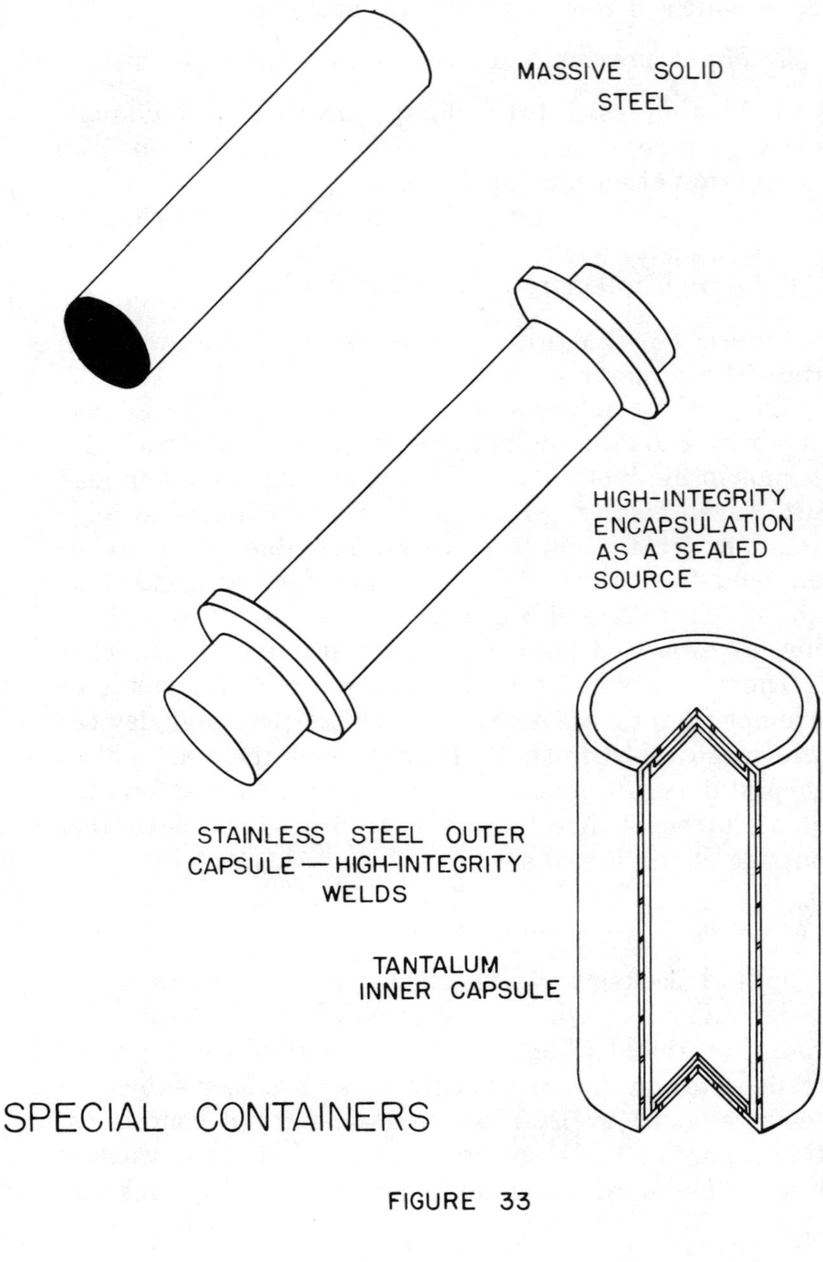
MASSIVE SOLID
STEEL
HIGH-INTEGRITY
ENCAPSULATION
AS A SEALED
SOURCE
STAINLESS STEEL OUTER
CAPSULE — HIGH-INTEGRITY
WELDS
TANTALUM
INNER CAPSULE
SPECIAL CONTAINERS

FIGURE 33

Table 5

Transport Group	**Type *A* Quantity**	**Type *B* Quantity***
I	Curies	Curies
I	0.001	20
II	0.05	20
III	3	200
IV	20	200
V	20	5,000
VI & VII	1,000	50,000
Special Form	20**	5,000

* Quantities exceeding Type *B* are large-quantity radiation sources. These and "Fissile Radioactive Materials" present special packaging problems. Their packaging is governed by regulations of the Department of Transportation and the U.S. Nuclear Regulatory Commission.

** For californium 252 the limit is 2 curies.

In addition, shippers may use their own containers after assessing them against performance specifications. Containers should be marked "Type A." Typical Type *A* packaging is illustrated in Figure 34.

Type *B* packaging must be designed to meet all Type *A* package requirements plus being able to withstand certain serious accident damage test conditions with only limited loss of shielding capability and essentially no loss of containment capability. Included among the package design test criteria are the following:

1. A 30-foot drop onto an unyielding surface.

2. A puncture test (a free drop of over 40 inches onto a 6-inch diameter steel pin).

3. Thermal exposure at 1475° F. for 30 minutes.

4. Water immersion for 8 hours (for fissile material packages only).

Typical Type *B* packaging is illustrated in Figure 35.

All Type *B* packages must be specification type packages or others whose design has been reviewed and approved by the Department of Transportation or other

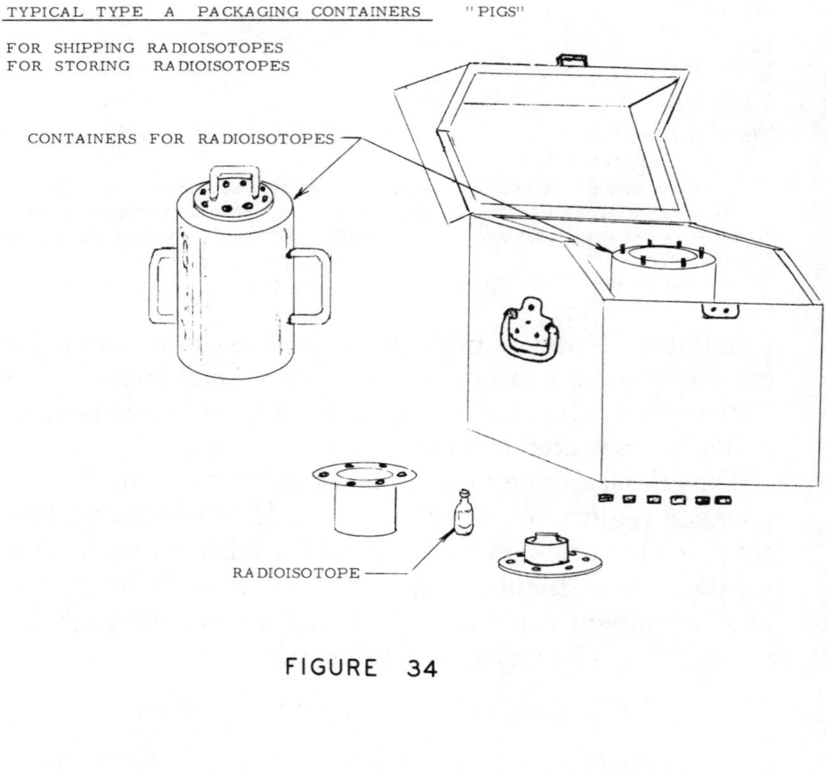

FIGURE 34

authorized federal agencies. Type *B* packages should be so identified.

Packaging of large quantities of radioactive materials (greater than permissible Type *B* allowances) require special safeguards over and above those required for Type *B*. Among the most important additional safeguards are:

TYPICAL TYPE B PACKAGING SCHEMES

PACKAGE MUST WITHSTAND BOTH NORMAL AND ACCIDENT TEST CONDITIONS WITHOUT LOSS OF CONTENTS.

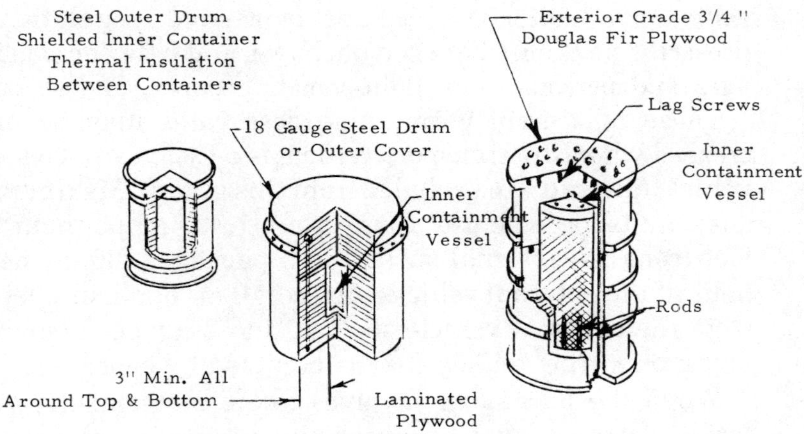

FIGURE 35

1. Dissipation of decay heat.

2. Control of potential leakage from contaminated heat transfer medium.

3. Heavier shielding

The most common commercial large quantity sources are high-curie irradiator sources and irradiated fuel elements.

While radioactive materials are in transit, current regulations set limits on the maximum permissible dose rate at point of contact with any accessible surface of a radioactive materials package. Rates are 200 millirem per hour at the package surface and a Transport Index of 10 millirem per hour 3 feet from the package. It should perhaps be pointed out that where the surface rate is 200 mr there will be shielding of some sort between the surface and a point 3 feet away. When more than one radioactive materials package is carried, regulations prescribe distances between packages and between packages and persons, film, light-sensitive paper, and so on.

These maximum permissible dose rates may be increased when the transport vehicle is a "sole use," closed vehicle (aircraft are excluded from this group). Maximum dose limits for "sole use" carriers are 1,000 mr per hour 3 feet from the external surface of a package, 200 mr per hour at the external vehicle surface, 10 mr per hour 6 feet from the external vehicle surface, and 2 mr per hour in any part of the vehicle that is occupied by a person.

When the packaging involves fissile materials, regulations are set up to prevent the possibility of fissile materials being present in such quantity and formation that the materials can "go critical" (begin a self-sustaining chain reaction). These additional requirements normally result in radiation dose levels being well below permissible limits.

Labeling

Every package of radioactive materials, except for those in exempt categories, must be labeled on two opposite sides by standard warning labels. Three such labels, shown in Figure 36, indicate the degree of hazard and provide information regarding the proper course of action in an emergency situation. The symbol with an all-white background indicates that the radiation level is minimal and no special action is needed (Class I). The symbol with a yellow background and two vertical red bars indicates a radiation level outside the package that may require special consideration (Class II). If the radiation level is still higher, there will be three red bars (Class III). Vehicles carrying Class III materials must be placarded. For both Class II and Class III materials, mandatory regulations bind both the shipper and the carrier. Required placards for rail and motor vehicles carrying Class III materials are shown in Figure 37.

When the radioactive materials shipped are of fissile type, the three classes become Fissile Class I, Fissile Class II, and Fissile Class III. The presence of fissile materials will be indicated on the markers.

Placards and labels by themselves show the degree of hazard but not the specific contents. For this reason it is important that a bill of lading or its equivalent be with the carrier. For rail shipments the documents will be with the freight train conductor and for motor vehicles with the driver or escort vehicles (many large shipments have escorts). These documents can be of vital importance in event of fire, accident, or other emergency.

Preparation by Firefighting Forces

As an additional safeguard firefighting forces should:

1. Have on hand railroad phone numbers to be called for emergency assistance.

2. Note the motor vehicle ownership and telephone number before they are obliterated by a fire so that proper information can be obtained from the shipper. (The driver could be incapacitated and the bill of lading destroyed.)

3. Unless the fire department has in-house capability, know where to get prompt radiological monitoring assistance.

AIR AND SURFACE

RADIOACTIVE MATERIALS LABELS

FIGURE 36-1

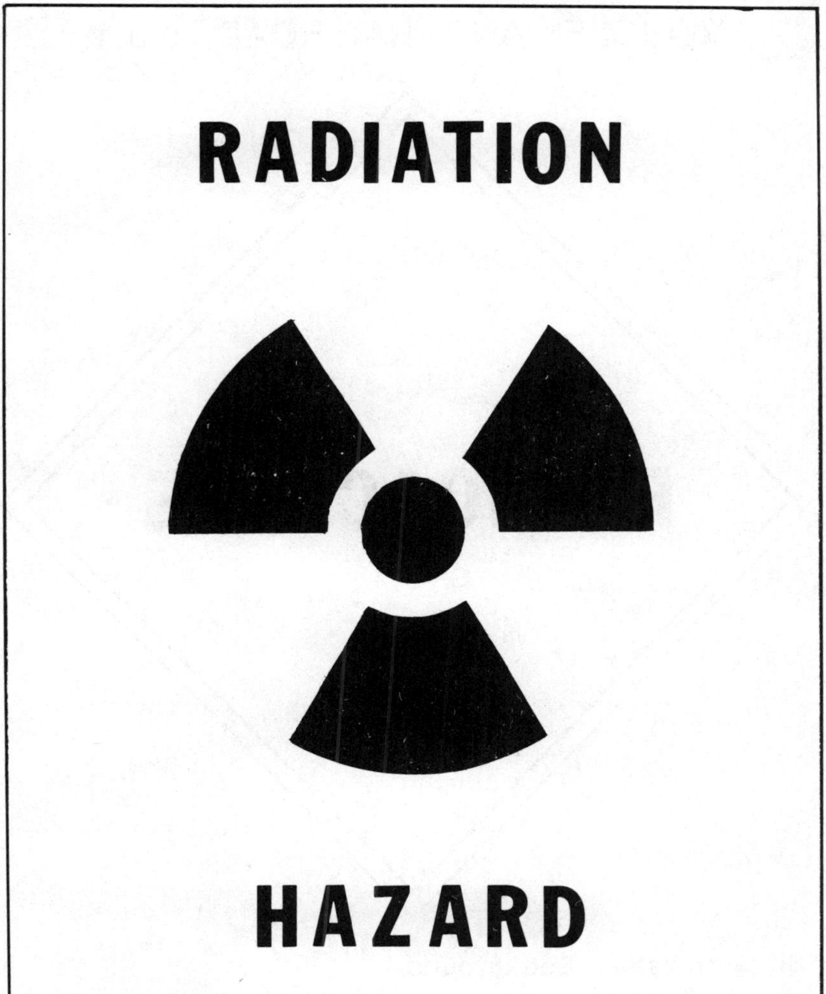

Black on Yellow Background

FIGURE 36-2

VEHICLE AND RAILROAD D.O.T.
PLACARD

Black on Yellow Background

FIGURE 37

4. Be familiar with the Interagency Radiological Assistance Plan (IRAP). Help can be obtained for radiation incidents, including fire and subsequent decontamination, from IRAP offices as shown in Table 6.

Large-Quantity Packaging

Before leaving the subject of packaging and labeling, let's look at the packaging for large-quantity radioactive materials. It is particularly important that fire and police department personnel learn to recognize these shipments. Figure 38 shows an irradiated fuel cask. Figure 39 shows an irradiated fuel cask on a flatcar (incidentally, rail cars carrying these casks should have roller or

Table 6
Interagency Radiological Assistance Plan Offices

Regional Area Coordinating Office	Address	Telephone
Brookhaven Area Office	Upton, L.I., N.Y. 11973	516-345-2200
Oakridge	P.O. Box E, Oak Ridge, Tenn. 37830	615-483-8611, ext. 3-4510
Savannah River	P.O. Box A, Aiken, S.C. 29801	803-824-6331, ext. 3333
Albuquerque	P.O. Box 5400, Albuquerque, N.M. 87115	505-264-4667
Chicago	9800 S. Cass Ave., Chicago, Ill. 60439	312-739-7711, ext. 2111 and 4451*
Idaho	P.O. Box 2108, Idaho Falls, Idaho 83401	208-526-0111, ext. 1515
San Francisco	2111 Bancroft Way, Berkeley, Calif. 94704	415-841-5121, ext. 664; 841-9244*
Richland	P.O. Box 550, Richland, Wash. 99352	509-942-7381

* Off-duty hours

Physicial Dimensions (Approximate)

Length	18 feet
Diameter	$5\frac{1}{2}$ feet
Weight	
empty	55 tons
loaded	67 tons

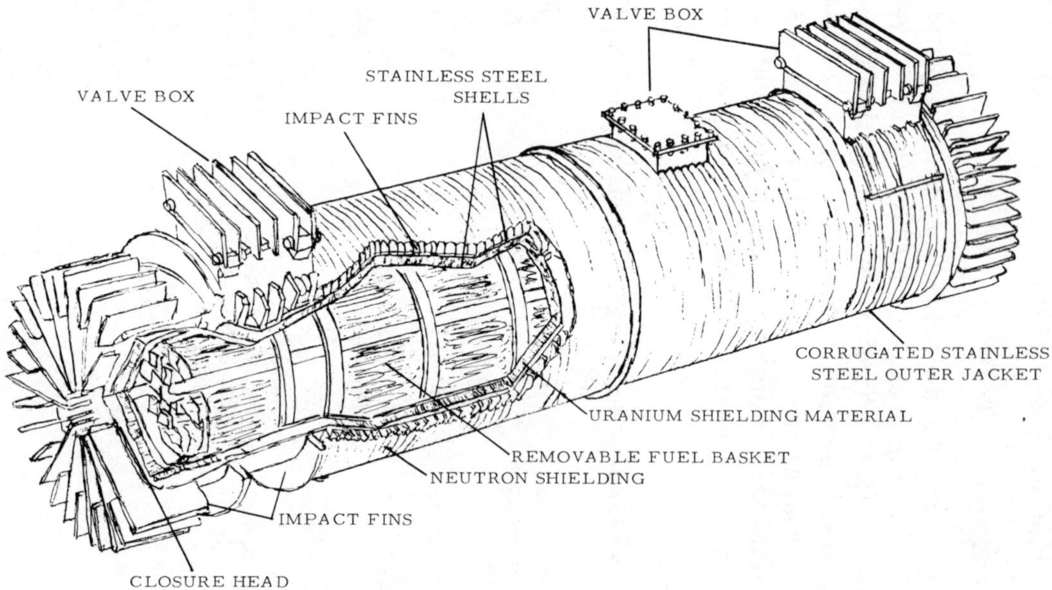

IRRADIATED FUEL CASK

FIGURE 38

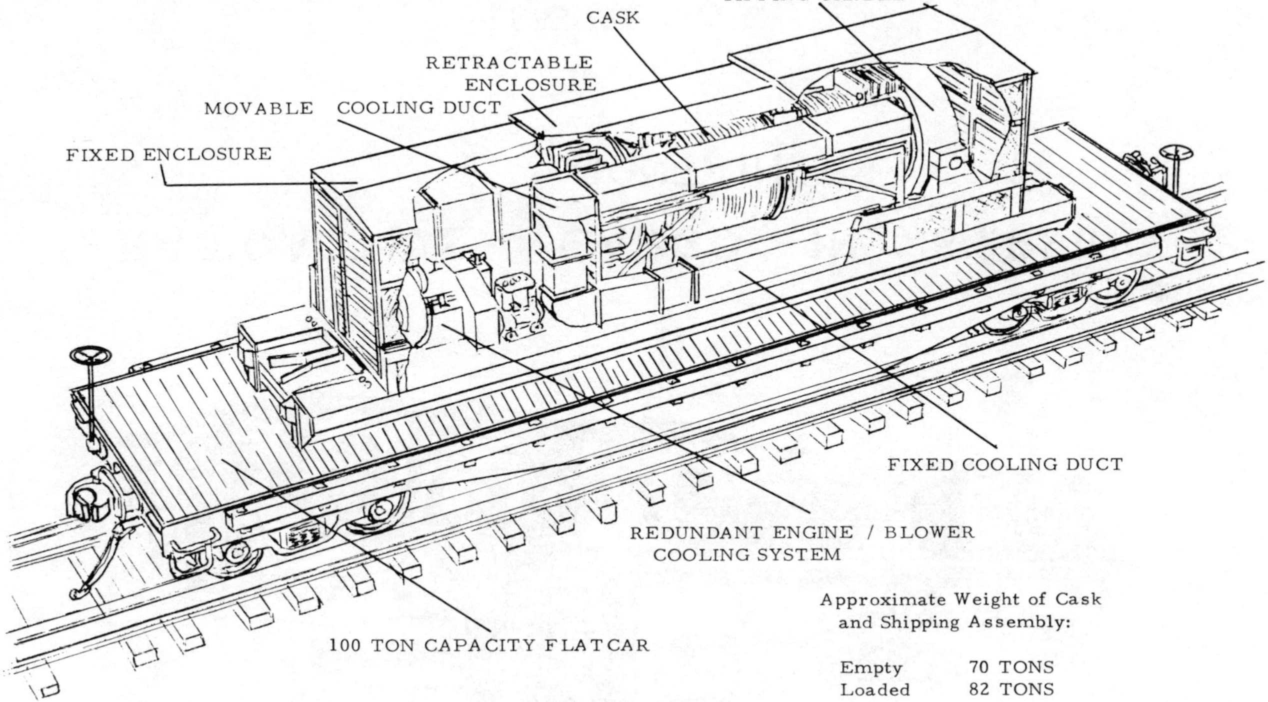

Approximate Weight of Cask
and Shipping Assembly:

Empty	70 TONS
Loaded	82 TONS

IRRADIATED FUEL CASK ON RAIL CAR

FIGURE 39

RADIOACTIVE CARGO IN SHIPPING CONTAINERS

FIGURE 40

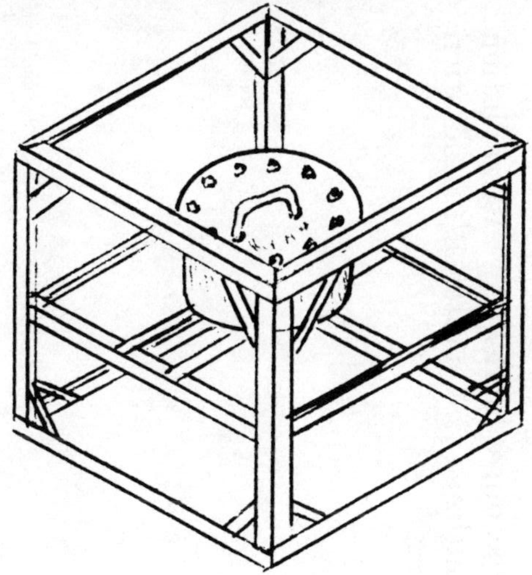

TYPICAL "BIRDCAGE" USED FOR SHIPPING FISSIONABLE MATERIALS

FIGURE 41

ball-bearing wheels and not journal boxes). Figure 40 shows a truck shipment. Figure 41 shows a "birdcage," typical packaging for fissionable material.

In checking actions to be taken in the event of fire or other emergency involving radioactive materials, it must be remembered that other hazards may also be present. It is entirely possible that at times the radioactive material or some other part of the shipment may be flammable, toxic, or corrosive, may react with water, and so on. All these factors must be taken into consideration in handling an emergency.

Basic firefighting procedures for dealing with radiation emergencies will be outlined in Chapter 17, "Radiation Hazards in Firefighting."

10

Radiation Instrumentation

Because radiation is odorless, colorless, and tasteless, it is necessary to have instruments to measure radiation levels wherever a harmful or potentially harmful amount of radioactivity may be present. If such instrumentation is not present, standard operating procedures should require all personnel to stay out of any radiation area where the radiation level is unknown.

Many types of instruments are used to detect radiation; the most common are Geiger counters, Geiger-Muller counters, film badges, and dosimeters. It must be remembered, however, that these instruments measure gamma and beta radiation only. If there is reason to believe that alpha radiation is present, it is necessary to use a special alpha meter. Aside from direct radiation exposure, there is a very limited possibility that in certain specific areas exposure to neutron bombardment could occur, and additional instruments measure the potential for this kind of exposure.

Basically, counters are used to record radiation levels for general use in making decisions, locating radioactive materials, and so on. Film badges and dosimeters record the exposures received by individuals.

Geiger and Geiger-Muller counters, in general, are relatively low-level (up to about 50 mr per hour) counters. Most of them measure both beta and gamma radiation. When the shield is off the tube, both types of radiation are recorded, and with the shield in place, only gamma radiation is recorded. The difference between the open and closed readings is the beta radiation level.

For higher or lower radiation readings ionization chambers and scintillation counters are used. The ionization chambers record high radiation levels. Scintillation counters are used to record precisely very low levels of gamma radiation.

While Geiger type counters are essential for making on-the-spot decisions in radiation areas, they provide no record of the exposures encountered by individuals in their daily work or in multiple emergency situations. Such recordings of the radiation doses absorbed by individuals are made by film badges and dosimeters. Both devices must be used in accordance with four cardinal rules if they are to serve their purpose.

1. Each device must be assigned to a specific individual.

2. Assigned devices must always be worn while in a radiation area and must be kept away from radiation when not being worn.

3. Exposure readings should be taken only by trained individuals.

4. A record must be kept of total dosage absorbed by each individual to see whether he reaches a point when he should no longer be allowed to go in a radiation area.

A film badge is simply a strip of film worn in a special holder. The film is sensitive to radiation in the same way that X-ray film is sensitive to X-rays. At appropriate times the film is developed and the degree of darkening on the film is compared with the darkening of films that have been exposed to known radiation sources. The film thus gives a picture of the radiation dosage the individual wearing it has received. When film badges are turned in for processing, it must be remembered that it takes time to develop and read the film; the worker should avoid additional exposures while the film is being processed and read.

Dosimeters come in several forms, the usual type for most fire safety personnel being the pocket dosimeter. This device, which looks like a ballpoint pen or an automatic pencil, may or may not be self-indicating. In either case readings should be taken by trained personnel. Dosimeters, if dropped or otherwise damaged, can give false readings. It is advisable, therefore, to wear two of them.

At locations where a sudden burst of high-level gamma radiation is a possibility, it is desirable to provide means for detecting it. Special instruments, which may be in fixed locations or worn on the person, will give an audible or visual alarm as appropriate whenever a predetermined radiation level is exceeded.

Alpha radiation, being extremely short-range, is detected by special alpha radiation meters. Each meter has a measuring surface that must be brought into contact with the contaminated surface. The nature of alpha radiation is such that the suspect area must be gone over inch by inch. For surfaces that are too irregular for contact, a designated area will be wiped with a cloth or a thin, absorbent paper and the radiation of the wiping material measured.

Some of the common radiation measuring devices are shown in Figure 42.

FIGURE 42

RADIATION MEASURING DEVICES

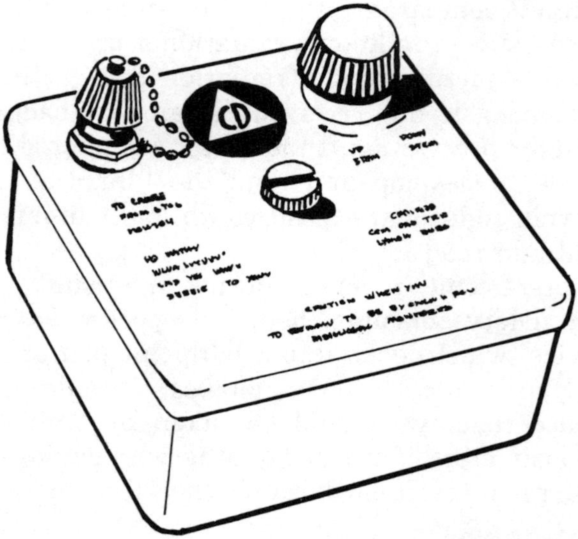

CD V-750 DOSIMETER CHARGER

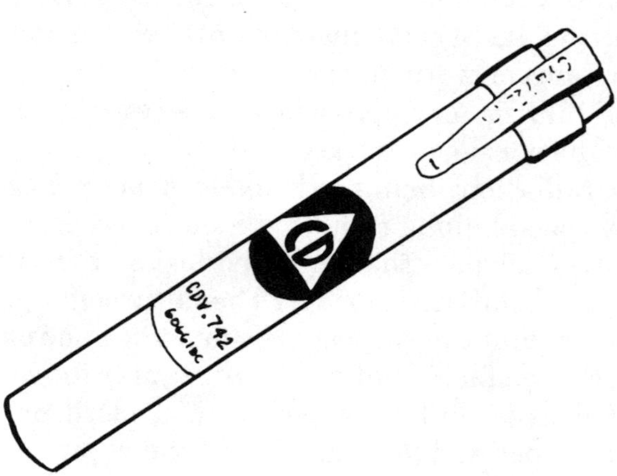

CD V-742 0-200r DOSIMETER

0–50 mr/hr SURVEY METER

0–500 r/hr SURVEY METER

11

Radiation Physics for the Layman

As previously noted, exposure to alpha, beta, or gamma radiation will not make a material radioactive, although, depending on the intensity of the exposure, the material will suffer varying degrees of damage. This holds true regardless of the location of the radiation source outside of or within the body.

It has also been noted that some materials can be made radioactive. It is further known that this radiation problem can be dealt with as successfully as can straight radiation problems. It is not necessary to be a nuclear physicist to handle this problem, but fire safety personnel do need to know:

1. How materials become radioactive.

2. The principles behind the generation of electric power from splitting the atom.

3. How fissile materials are handled.

4. The difference between fissile and fertile materials.
5. How radioactive materials decay.

Atomic Structure

The logical starting point for fire safety personnel is to investigate the structure of the atom in somewhat more detail than was done in earlier chapters. The atom as a whole behaves as if it consists of an extremely dense, hard nucleus about which electrons orbit in about the same manner as the planets of the solar system orbit the sun. Most of the atom's weight is in the nucleus, the electrons having only insignificant weight. Other subatomic particles also exist, but these do not affect radiation problems with which fire safety personnel will be concerned. These particles will, therefore, be disregarded.

The nucleus is made up of tiny particles of matter, tightly bound together. Some of these particles or balls of matter, as they are sometimes called, carry a positive electrical charge; others are electrically neutral. The charged particles are protons and the electrically neutral particles are neutrons.

The electrons orbiting the nucleus are negatively charged. Normally, an atom as a whole is electrically neutral, meaning that for each proton in the nucleus there is an electron with an equal and opposite charge.

Chemically speaking, the number of protons in the nucleus determines what the element is. If the number of protons in the nucleus changes, the element also changes. For example, uranium atoms have 92 protons. If ten protons are removed from each uranium atom, the result is lead. There are somewhat over one hundred elements. Those of interest from the radiation standpoint to fire safety personnel range from hydrogen with one proton per atom, to uranium with 92 protons, to manmade plutonium with 94 protons. The heavier elements, found in nature or manmade, are in too small quantities or too short-lived to affect the radiation hazard fire picture. It is interesting to note at this point that the reason medieval alchemists failed to convert one element to another was that they used *chemical* means. Recognizing that it is the *physical* structure of the atom that determines its chemical properties, modern scientists have succeeded in converting one element to another. It is an expensive process, but it can be done rather easily.

Returning to the nucleus for a moment, different nuclei of the same element all have the same number of protons. In certain elements, however, the number of neutrons per nucleus varies. Depending on the element, the number of neutrons in the nucleus can vary from none to approximately 150. These variations in the number of neutrons affect some physical properties and in unstable atoms enter into the radiation problem. Figure 43 illustrates the arrangement of a typical lithium atom.

Atoms with differing numbers of neutrons in their nuclei have different weights, even if they are atoms of the same element. These variations are called isotopes. Most isotopes are stable, but some are not. These unstable or radioactive atoms are of major concern to both nuclear physicists and fire safety personnel, who may have to deal with them in industry or on the fireground.

Atomic Weight and Atomic Number

The question of isotopes illustrates the distinction between the terms *atomic number* and *atomic weight.* The atomic number of an element is the number of protons in the nucleus. The atomic weight is the sum of the number of protons and the number of neutrons in the nucleus. Obviously, therefore, atoms of the same element with varying numbers of neutrons per atom will have different

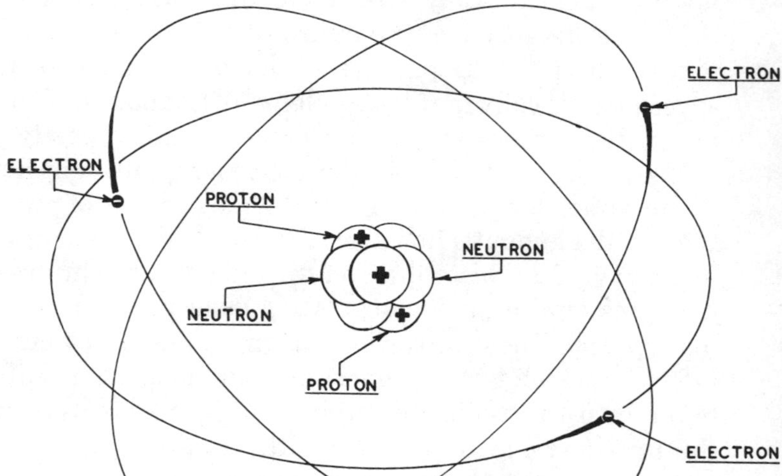

THE WORKING PARTS OF A LITHIUM ATOM - ELECTRONS, PROTONS, & NEUTRONS

FIGURE 43

weights. This variation will not show up in the usual chemical tables, which generally list an average weight. For example:

1. Hydrogen occurs in three forms or isotopes as follows:

	Protons	Neutrons	Atomic Number	Atomic Weight
Hydrogen	1	0	1	1
Deuterium	1	1	1	2
Tritium (radioactive)	1	2	1	3

Based on the percentage of each isotope, the average atomic weight is 1.008, which is the figure used in chemical tables.

2. Uranium occurs naturally in several isotopes of which only U238 and U235 are of importance to fire safety personnel.

	Protons	Neutrons	Atomic Number	Atomic Weight
Uranium 238	92	146	92	238
Uranium 235	92	143	92	235

Average of all isotopes in natural uranium is 238.03

Perhaps the easiest way to illustrate the possible effect of variations in the number of neutrons is to analyze water. Normal water consists of two hydrogen atoms and one oxygen atom per molecule. The molecular weight is approximately 18. If deuterium, which chemically is hydrogen, is substituted for normal hydrogen, the result is heavy water with a molecular weight of 20. If tritium is substituted for normal hydrogen, the result is radioactive heavy water with a molecular weight of 22. All three variations look alike, taste alike, and react in the same way chemically, but the last two types would be dangerous to drink or use in many ways in which normal water is used. As a matter of interest, during World War II the Nazis were experimenting with heavy water as one path toward atomic weapons.

Decay of Radioactive Atoms

What happens when a radioactive element, isotope, or atom decays? Basically a disintegrating atom will emit alpha and/or beta particles until what remains of the original atom is an atom of a stable element. For example, a uranium 238 atom will decay step by step until it becomes lead. At that point disintegration stops. At some steps in the disintegration process, gamma radiation is also thrown off. This occurs when a particle departing from the nucleus does not take with it sufficient energy. This surplus energy is thrown off as gamma radiation. Omitting the problem of gamma radiation, the decay of a uranium 238 atom is shown in Table 7.

It should be noted from this table that:

1. There are two uranium isotopes. They are chemically identical.

2. There are two bismuth isotopes. They are chemically identical.

3. There are three lead isotopes. All three are identical chemically, but two isotopes are radioactive, and one is not.

In an earlier chapter it was noted that an alpha particle is two protons and two neutrons (the nucleus of a helium atom) which are ejected together from the core of the radioactive element. It has also been noted that a beta particle is an electron. A beta particle is formed when a neutron breaks up into its component parts, which are one proton and one electron held together by a binding energy. When this energy is released, the electron is ejected as a beta particle as illustrated in Figure 44.

There is no way to determine at what point in time an individual atom will disintegrate. In the case of a large number of atoms, however, it is possible to determine how long it will take for half the atoms to disintegrate. This is the half-life of the radioactive material and, depending on the element, may be anywhere from less than

Table 7

Element	Atomic Weight	Type of Radiation	No. of Protons	No. of Neutrons
Uranium	238	Alpha	92	146
			−2	−2
Thorium	234	Beta	90	144
			+1	−1
Protactinium	234	Beta	91	143
			+1	−1
Uranium	234	Alpha	92	142
			−2	−2
Thorium	230	Alpha	90	140
			−2	−2
Radium	226	Alpha	88	138
			−2	−2
Radon	222	Alpha	86	136
			−2	−2
Polonium	218	Alpha	84	134
			−2	−2
Lead	214	Beta	82	132
			+1	−1
Bismuth	214	Beta	83	131
			+1	−1
Polonium	214	Alpha	84	130
			−2	−2
Lead	210	Beta	82	128
			+1	−1
Bismuth	210	Beta	83	127
			+1	−1
Polonium	210	Alpha	84	126
			−2	−2
Lead	206	Stable Element	82	124

a microsecond to thousands of years. The half-life is important in determining whether or not the material is usable and, if so, for what; how the material must be transported; and how waste materials must be disposed of.

For material such as uranium 238, which must go through a number of disintegrations to become the stable

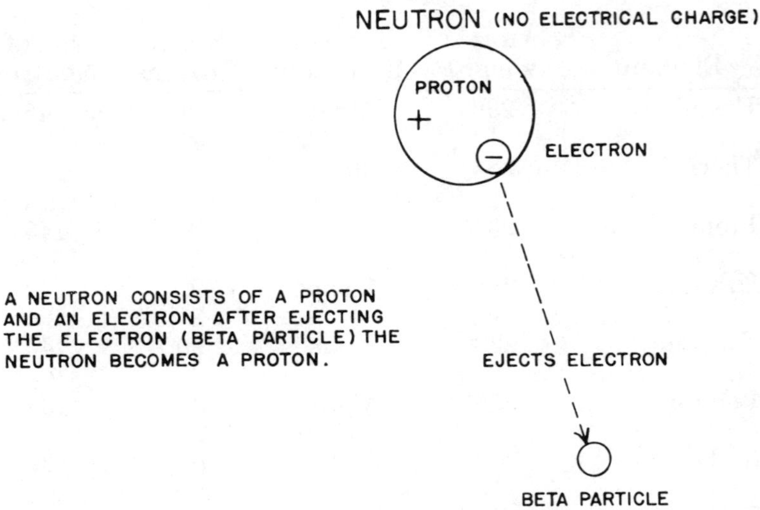

FIGURE 44

element lead, it is unsafe to assume that only one type of radiation is being given off. It is possible that all three types of radiation may be given off at the same time. Proper instrumentation and monitoring where radioactive materials may be present in quantity are essential to provide accurate information about the types and amounts of any significant radiation.

Atomic Fission

If radiation physics were limited to naturally radioactive materials and isotopes, the area of concern would be very small. In fact, most of the beneficial uses mentioned in Chapter 2 would not exist. For example, cobalt 60 treatment for cancer would be impossible, as cobalt is not naturally radioactive. For another example, there would be no possibility of producing electric power on a large scale. Only about fifty radioisotopes occur naturally, whereas there are well over 1,000 artificially produced radioisotopes. To understand the production of these extra radioisotopes and the capability of producing energy in large amounts it is necessary to investigate how fissionable and fertile materials act.

What is atomic fission? How is usable energy in large amounts obtained by splitting or fissioning an atom? The simplified explanation in the following paragraphs provides the data essential to fire safety personnel.

It can be mathematically proven that mass can be converted into energy. For example, if a uranium atom is broken into two smaller atoms, the combined weight of the two small atoms is less than the weight of the original atom, and a measurable amount of energy has been released. The loss in weight produced the energy.

To break up (fission) a nucleus, an atomic particle that has sufficient energy and mass to penetrate the nucleus is necessary. Except under very special conditions, alpha, beta, and gamma radiation are not capable of breaking up a nucleus, but neutrons are. Neutrons, being electrically neutral, are not repelled by the positive charge of a nucleus; they can enter a nucleus with no opposition.

While the process just described is complete in itself for most radioactive materials, something more is required to produce sufficient energy to provide a useful power supply. This "something more" is an element that is fissionable and that can be arranged to set up a continuing chain reaction.

In the natural state only the U235 isotope of uranium is both fissionable and present in sufficient quantity to be useful for energy production. When a U235 atom is hit by a neutron and split into smaller particles, some of the neutrons from the original nucleus become free agents. Some of these fast-moving, free neutrons strike other nuclei of the same material, splitting them and thereby creating more energy and more free neutrons to continue the action. This is the secret behind the release of large quantities of energy from the atom.

Before proceeding further, it is essential to emphasize that the terms *radioactive* and *fissionable* are not interchangeable. Most radioactive materials are not fissionable and could not participate in a chain reaction. As noted in the previous paragraph, only U235 (0.07 percent of all natural uranium) is readily available in the natural state; all other fissionable materials are manmade from *fertile* materials.

A radioactive material is called *fertile* when it is capable of being converted into fissionable material by the capture of a neutron. Uranium 238 and thorium 232 are examples of fertile materials. Manmade fissionable materials in common use are plutonium 239 and uranium 233 (there are others). These materials are generally made in breeder reactors.

In theory, since radioactive isotopes, capable of fissioning, are in the process of disintegrating all the time, it is possible for a naturally occurring fissionable material to undergo a chain reaction. Under natural conditions the possibility of an important chain reaction occurring is extremely remote due to the small quantity of fissionable material in any one place. The process of creating an important chain reaction is about as follows, starting with a small amount of U235.

1. The material fissions and throws out some free neutrons, some of which stay in the mass and more of which escape into space. The free neutrons remaining in the mass cause more atoms to fission, but as long as more neutrons escape than remain there will be no self-sustaining chain reaction. There will be a series of tiny chain reactions that will be absorbed or "fizzle out" before they become self-sustaining. This amount of material is called a *subcritical mass*.

2. As additional subcritical masses are added to the original mass, a change begins to take place if the rate

of increase of the mass is greater than the rate of increase of the surface area. The percentage of free neutrons remaining in the mass increases and the percentage escaping decreases. Eventually a point is reached where the number of free neutrons produced equals the number escaping. At this point the mass is ready to go critical. When the number of neutrons remaining equals the number escaping plus one, the mass goes critical and a self-sustaining chain reaction begins.

3. During a chain reaction, release of energy results in a rapid heat buildup.

The material then softens or melts and spreads out, providing more surface area in proportion to the mass. As the mass spreads out, more neutrons escape and fewer remain in the mass. When the number escaping exceeds the number remaining, the chain reaction stops.

This process, assuming an unconfined mass, takes only a few moments. However, while it is going on there will be:

1. An emission of gamma radiation which would seriously expose people in the area.

2. A neutron bombardment which would be hazardous to persons present. If sufficiently intense, personnel and items exposed could become radioactive.

3. The evolution of considerable heat.

4. The production of various types of fission products which in turn would give off different types of radiation and have widely varying half-lives.

5. A distinct possibility of contamination of the immediate area, any ventilating system, and other areas served by the ventilating system. (Normal air currents

induced by the air handling system and convection currents created by heat emission could readily spread many types of fission products.)

Anyone in the area when the mass went critical would be injured or possibly killed, depending on his proximity to the critical system. Persons entering the area later would have to be suitably protected against possible radiation exposure from the fissioned materials. There would not, however, be an atomic explosion. Without the

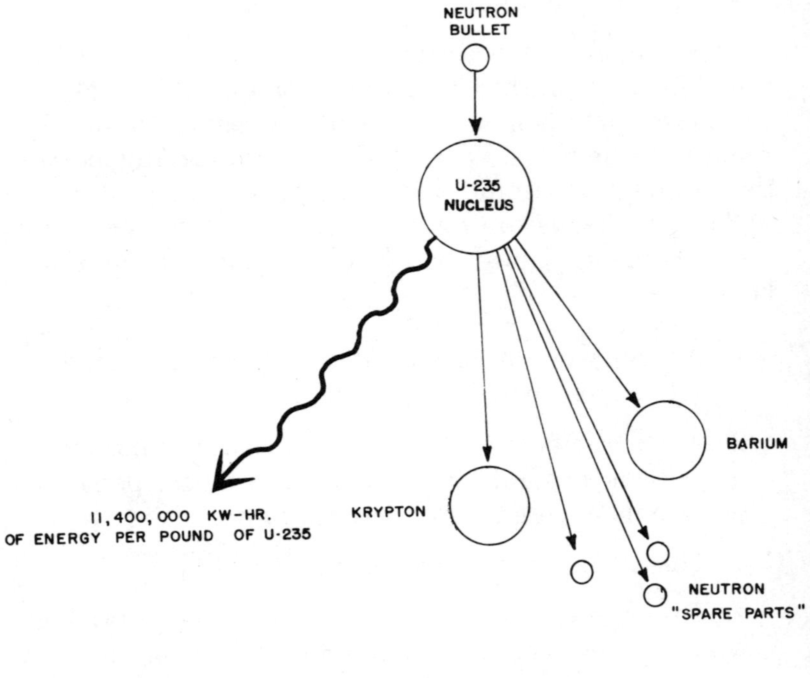

FIGURE 45

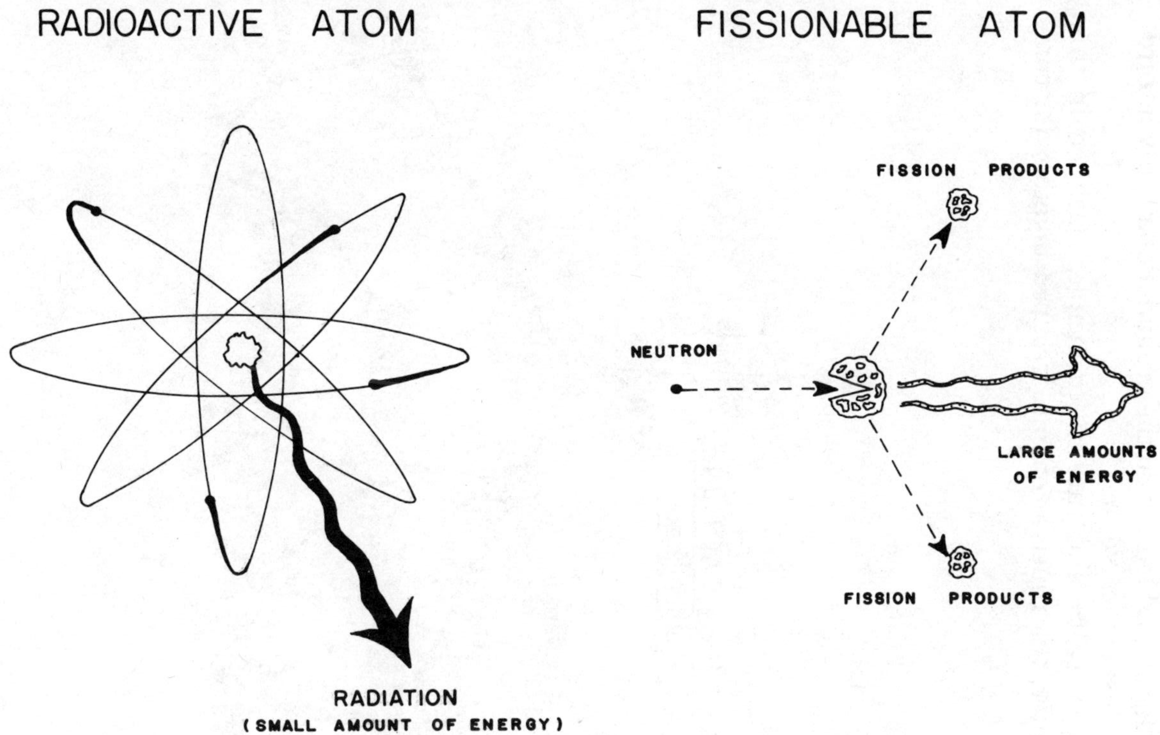
RADIOACTIVE ATOM
RADIATION
(SMALL AMOUNT OF ENERGY)
FISSIONABLE ATOM
NEUTRON
FISSION PRODUCTS
LARGE AMOUNTS OF ENERGY
FISSION PRODUCTS

FIGURE 46

necessary hardware, an explosion from a chain reaction cannot take place any more than it could from free black powder spread loosely on the ground. Potentially hazardous, yes; explosive, no.

In order to illustrate the principles outlined in this chapter, the illustrations are grouped together for com-

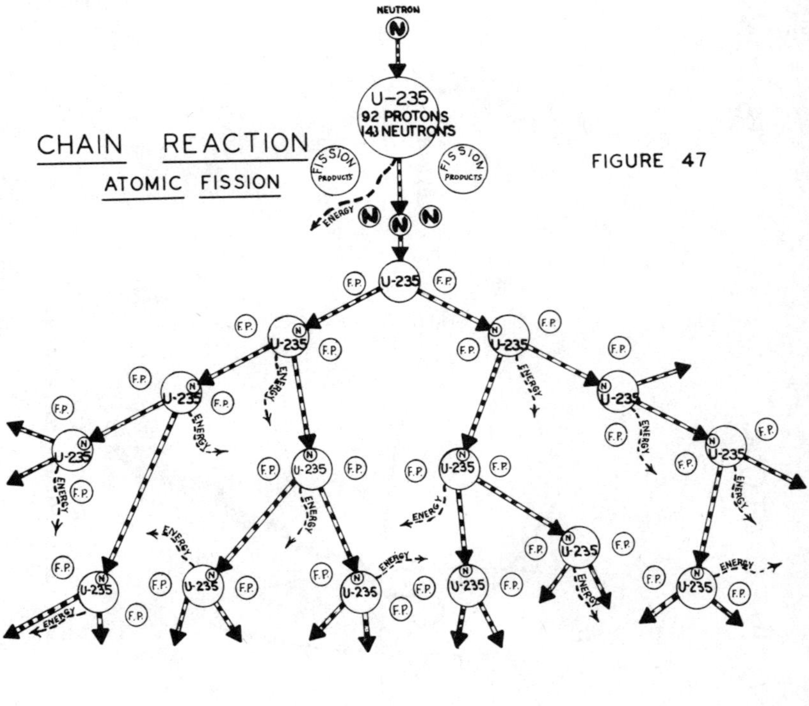

FIGURE 47

parison purposes. Figure 45 shows the energy production from the decay of a radioactive atom. Figure 46 shows the energy production resulting from an atom fissioning. Figure 47 shows the development of a chain reaction. Figure 48 shows the critical mass concept.

To gain any real value from the properties of fission-

CRITICAL MASS CONCEPT
FIGURE 48

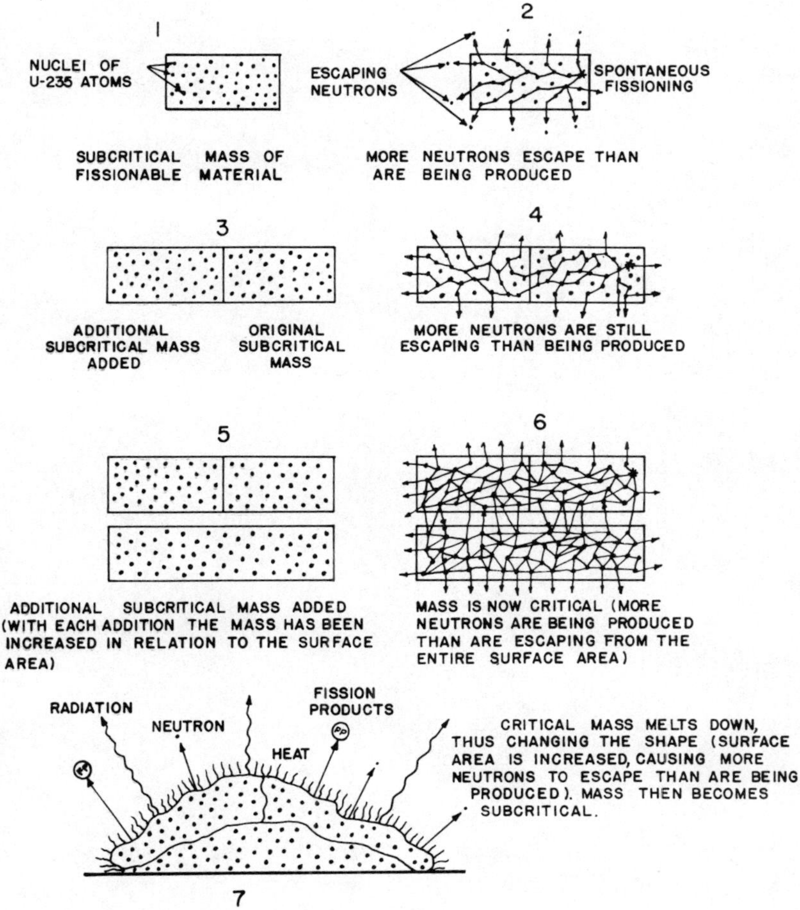

able material, it is necessary to put these properties to work in controlled situations. In Chapter 12, "Nuclear Power Plants," and Chapter 13, "Radiation Machinery," the application of these principles to the production of power and radioisotopes and the transmutation of elements will be outlined from the standpoint of fire prevention personnel.

12

Nuclear Power Plants

Although fire safety personnel need not understand the detailed design of nuclear power plants, they should be aware of the principal components of nuclear reactors and of a few key operating principles.* Behind these principles is what should be a mandatory axiom: safe control of sustained chain reactions.

Reactor Components

The principal components of nuclear reactors are:

1. *Reactor Fuel.* Many types of fuels are used. While some experimental fuels have been liquid, all commercial fuels are solid, usually in the form of strips, rods, or plates (so-called radioactive cores).

2. *Blankets.* In breeder reactors blankets of fertile ma-

* Given the rapid development of nuclear technology, changes in the materials used may occur at any time. Such advances do not change the basic principles described here, however.

terial are placed in or around the reactor core for the purpose of creating fissionable material.

3. *Moderators* (materials of low atomic weight which have the ability to slow down fast neutrons without capturing too many of them). Common moderators include water, heavy water, and carbon.

4. *Reflectors*. Reflectors surround the fuel cores to prevent the escape of too many neutrons.

5. *Reactor Vessels*. These are pressure vessels that (depending on size or use) may operate at anywhere from a few pounds per square inch (psi) and 300°F. to well over 2,000 psi and 1,000°F.

6. *Thermal Shields*. Usually these are steel plates around the reactor core to conserve heat and reduce stresses in the reactor vessel proper. Such shields may also have a circulating cooling water system.

7. *Biological Shields*. These shields, which are usually dense concrete or iron, are placed around the reactors to protect personnel from radiation.

8. *Cooling Systems*. All reactors produce heat, which must be used or dissipated. While heat from smaller, nonpower reactors may be dissipated by radiation, convection, or conduction, the larger reactors and all power reactors require complete cooling systems, including circulation piping, pumps, heat exchangers, and other equipment as appropriate. The most common cooling systems use water or gases. Liquid sodium and liquid sodium-potassium have also been used. If operations are to be safe, there must be a continuous and adequate supply of coolant. Frequently, to assure maintenance of cooling capacity, backup cooling systems are provided. Loss of cooling capacity during operating periods can result in a core melt-down with a consequent potential for serious damage. It should also

be borne in mind that liquid sodium or liquid alloys of sodium and potassium will present a special fire problem if the closed cooling system should fail.

9. *Reactor Controls.* Various materials are used for controlling nuclear reactions; all have a high affinity for neutrons and work by drawing neutrons away from the chain reaction. Of these, by far the most common are boron, cadmium, and hafnium fabricated into plates or rods. To assure reliability, more than one means should be provided to move these rods or plates into the reactor.

Principles of Reactor Operation

Using the components listed above and the facts about radiation that were developed in Chapter 11, it is quite easy to develop the basic principles behind a simple atomic reactor. Principal steps to be taken are:

1. Assume two subcritical masses of U235, U233, or Pu239 which, if brought close together, would go critical.

2. Insert boron rods between the two masses.

3. Bring the two masses close together but with the boron rod completely separating them. Both masses remain subcritical.

4. Withdraw the rods all the way and the two masses immediately go critical.

5. Withdraw the boron rods only part way, to a point where only the proper number of neutrons required to keep the action going at the desired rate would be produced.

6. The controlled chain reaction produces heat which, if not removed, would cause a core melt-down. In a large reactor such a melt-down could have serious con-

sequences. To prevent a melt-down, a circulating heat transfer system is installed. Transfer medium is usually water. Liquid sodium has been used in demonstration units and will be the medium used in fast breeder reactors.

7. Piping from the heat transfer system passes through a heat exchanger in which water is converted to steam. This steam is used to power a steam turbine which in turn drives an electric generator.

8. Condensate from the steam turbine passes through a condenser and thence back to the heat exchanger to repeat the process.

These steps occur in a simple atomic reactor capable of producing electric power. In effect such a reactor would power a small nuclear power plant. Figure 49 illustrates one concept of a simple atomic reactor.

Control Systems

It is possible for something to go wrong at a reactor (for example, loss of the heat transfer medium). If this occurs the reactor is "scrammed" by shoving the boron rods all the way in. This stops the chain reaction and the production of heat. If this system should fail, the potential for serious damage from a runaway reactor would escalate rapidly as reactor size increased. (Damage would be from radiation and not an atomic explosion.) It is, therefore, vitally important that primary controls for "scramming" a reactor be kept in good operating condition at all times. To guard against the slight possibility of control failure (such as inaccessibility because of fire or other incident) backup means should be provided for direct or remote operation of control rods.

A number of nuclear reactors now in use or proposed are considerably larger and more complex than the simple unit illustrated. Some reactors use different fuels,

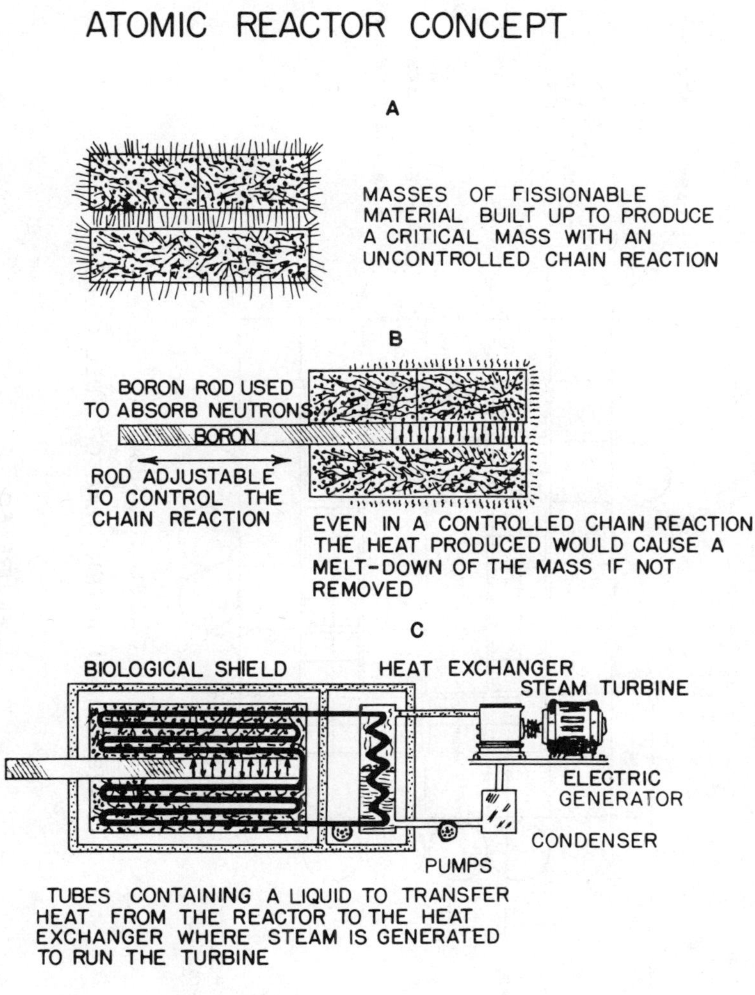

FIGURE 49

different types of control systems, different cooling systems, and so on, but basic operating principles will remain essentially the same. Schematic diagrams of key reactor concepts are shown in Figures 50–52.

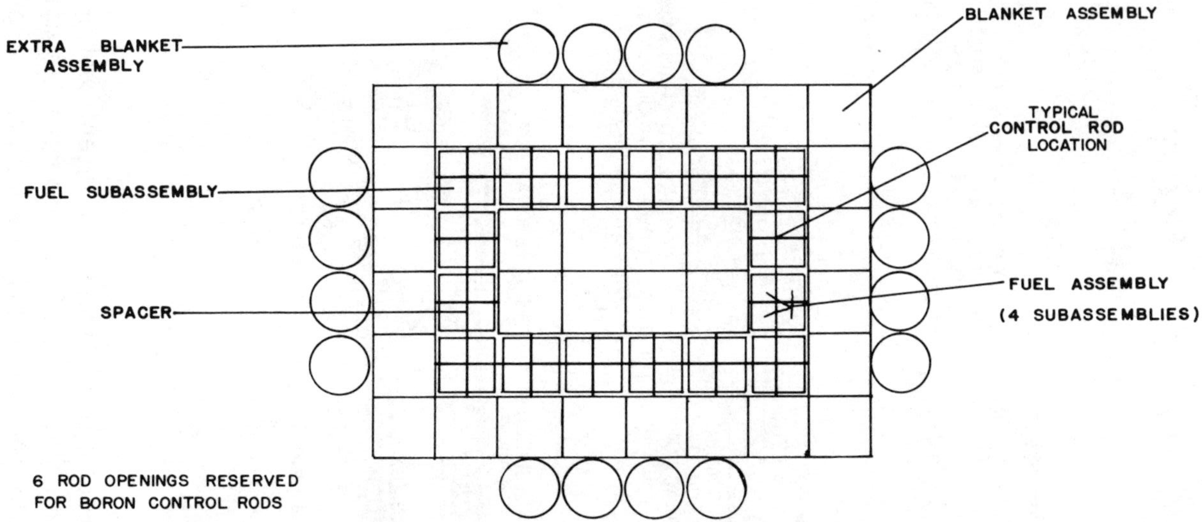

CROSS SECTION

PRESSURIZED WATER REACTOR FUEL CORE

FIGURE 50

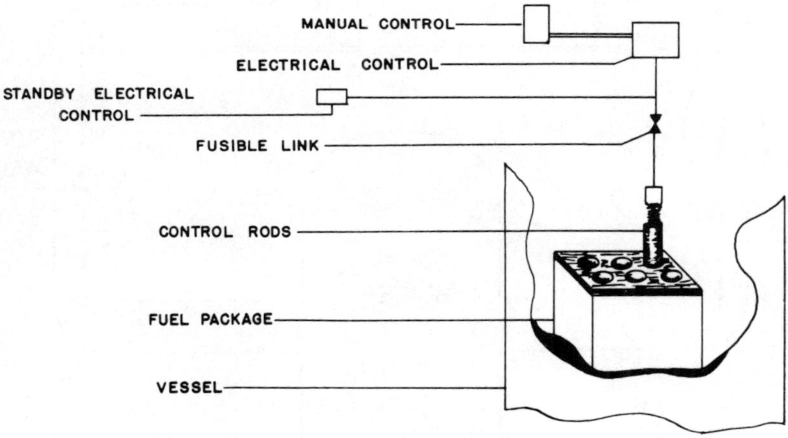

BORON ROD CONTROL

FIGURE 51

Production of Radioisotopes

Production of power, however, is not the only useful function of a reactor. A reactor also can be used to produce radioisotopes. Assume for illustration purposes that

FIGURE 52

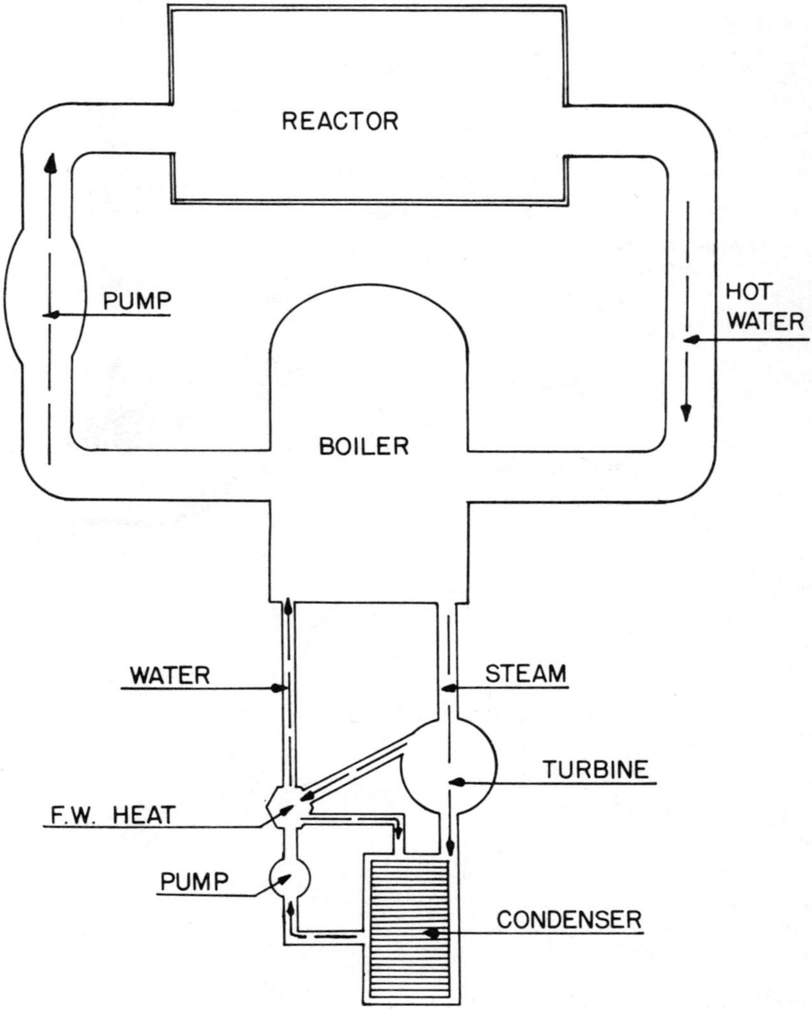

PRESSURIZED WATER REACTOR WATER-STEAM CYCLE

an operating reactor is designed with seven rod openings into the radioactive core and that only six of these openings are required for boron control rods. The seventh opening is, therefore, available for the insertion of other material.

In the seventh opening, the space between the fissioning masses is filled with neutrons entering and leaving their respective masses. This neutron movement, which would subject any material entering the space to an intense bombardment, is called a *neutron flux*.

To simplify the problem, assume a cobalt rod is inserted into the neutron flux. Natural cobalt has only one form: a nucleus with 27 protons and 32 neutrons plus 27 electrons. While the cobalt 59 rod is in the neutron flux a few of the cobalt atoms will capture one neutron each. These atoms are no longer natural cobalt but have become cobalt 60, which is radioactive, gives off gamma radiation, and has a half-life of 5.24 years.

Although only a small part of the cobalt rod becomes radioactive by this process, there is no practical way to separate out the radioactive cobalt. Despite the small size of the radioactive portion, however, cobalt 60 is valuable as a tool for treating cancer, detecting metal flaws, and so on.

The process just outlined is the method used to create many artificial radioisotopes.

As another example of what can be done in a reactor, let's substitute mercury 200 (a common mercury isotope) for cobalt. Mercury 200 nuclei have 80 protons and 120 neutrons. In the neutron flux field a tiny fraction of the mercury atoms each captures a neutron. These atoms become mercury 201, which has 80 protons and 121 neutrons in each atom. Unlike cobalt 60 and certain other elements, mercury 201 in a neutron flux becomes excited, and each excited atom throws out a proton. Each nucleus then has 79 protons and 121 neutrons. While the atomic weight is identical with that of mercury 200, the element is no longer mercury but gold 200 (a radioactive form of gold). Since a different element has been created, it is possible to separate the gold by chemical means. This type of process makes a material radioactive while simultaneously transmuting it into another element.

Figures 53 and 54 illustrate how reactors are used to make materials radioactive.

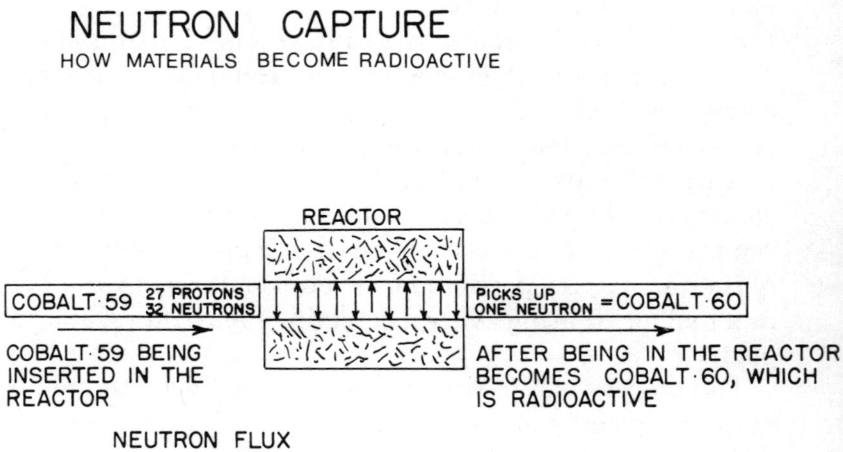

FIGURE 53

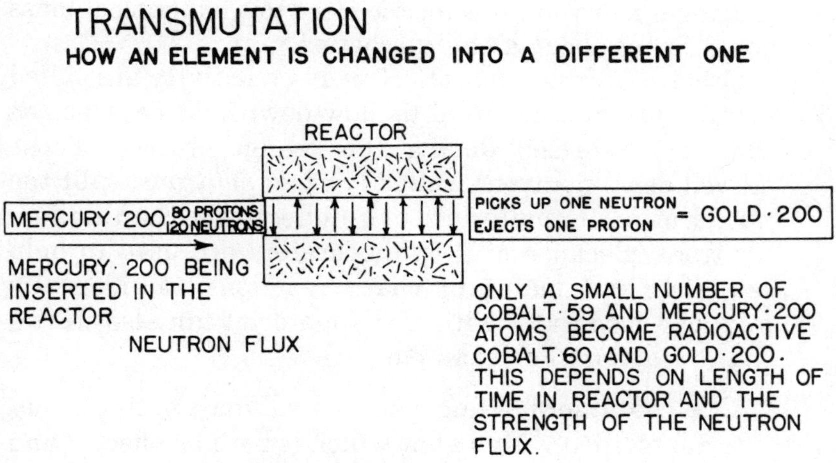

FIGURE 54

In sum, radioisotopes can be obtained from reactor operation in three basic ways: (1) gathering up and separating out fission products, (2) neutron capture, and (3) neutron capture and transmutation. Materials are not made radioactive by exposure to alpha, beta, or gamma radiation as normally used; they are made radioactive by being subjected to an intense neutron flux.

Improving Reactor Efficiency

To improve the efficiency of a reactor, neutrons must be conserved. To accomplish this, a material that will not absorb many neutrons and is capable of reflecting free neutrons back into the core is needed. Such materials are called *reflectors;* examples are some light elements and materials containing considerable hydrogen. Accidentally placing a neutron reflector too close to a subcritical mass could bring about a chain reaction.

Other materials that affect neutron activity are called *moderators* and are used to slow down fast neutrons as they are released during the fission process. These slowed-down neutrons, called *thermal neutrons,* split the nuclei of fissile materials more effectively than do fast neutrons. Because of this, moderators are used to help establish and maintain chain reactions. Accidentally mixing a moderator with fissile material can also start a chain reaction. For example:

1. A container holding a subcritical mass of fissile material is filled with water, which is both a reflector and a moderator. The mass goes critical.

2. A container holding a subcritical mass of fissile material is placed in a slightly larger container with water-jacketed walls. Again, the mass goes critical.

Fire and Explosion Safety

What has all this to do with safety? It must be borne in mind that all three deaths in the radiation field (cited in Chapter 4) occurred when subcritical masses accidentally became critical. These incidents could be repeated in event of fire near a reactor, unless firefighting forces know the potential hazard and have preplanned their actions accordingly. It must be remembered that some fire extinguishing agents have the same properties as materials used as reflectors and moderators. This does not mean that these tools cannot be used. It does mean that they must be used properly.

Before leaving the subject of nuclear reactor components and their arrangement, one additional factor requires mention. Even in the most efficient reactor some neutrons escape through the reflector and get lost to the process. If enough of them escape, there will be a personnel hazard. Accordingly, a biological shield, commonly made of dense concrete, is provided to absorb escaped neutrons. The integrity of this shield must be maintained, for cracks in it may be dangerous.

Based on an understanding of the basic principles on which nuclear reactors function, fire safety personnel should take a good, hard look at the principal fire and explosion hazards that need to be guarded against.

1. Light and heavy water (reflectors, moderators) are decomposed by the intense neutron bombardment to which they are subjected. Hydrogen, deuterium, and oxygen in hazardous combining quantities are released. Reactors must be designed to prevent these gases from combining, except for those types of reactors that have been designed to include this combining process as part of the operation.

2. The light metals and metal alloys such as sodium, potassium, and sodium-potassium alloy that are used for heat transfer mediums are combustible in air and ignite spontaneously in contact with moisture. These metals are in closed systems that should be kept closed.

3. While unlikely, mechanical rupture of a reactor is possible. In such an event considerable radiation would be released. Radiation levels outside the reactor would in all probability make the building or housing area untenable.

4. In reactors using water for moderators or for cooling, loss of the water would cause severe damage if the

reactor was operating at or near full power at the time. Severe damage could also occur if the water moderator was withdrawn or the cooling system stopped too soon after a shutdown. Reactors need to be cooled slowly so that short-lived radioisotopes have time to lose most of their radioactivity. Under extreme conditions, it would be possible for the radioactive materials to become airborne and spread radiation and fission products over a wide area.

5. If mistakes are made during a fire, an accidental assembly of a critical mass could occur. There would then be an uncontrolled nuclear reaction. It would not be an atom bomb explosion, but widespread contamination with consequent personnel hazard could result.

6. Fire in or near the control panel could have very serious consequences, unless other means are provided to operate the rods or other standby controls so that the reactor can be shut down. One possible consequence of control failure is reactor rupture.

7. Neglect of the ABCs of fire prevention and protection has been the reason for many large fire losses in all types of property. Where reactor facilities are concerned, overlooking these fundamentals could be extremely serious because of the presence of radioactive and fissile materials. Again, the danger is not from an atom bomb type explosion, but from widespread contamination. In addition to principles of fire resistive or noncombustible construction, fundamentals include but are not limited to:

a. Wiring. Since it is possible to get almost every type of wiring with noncombustible or flame retardant insulation, the use of wiring serving reactor controls should not as a rule involve combustible insulation. Where combustible insulation does exist, important cables should be made flameproof. Separation of important cables and/or groups of cables so that fire or a short circuit in one cable will not affect a large number of cables or the ability to operate reactor controls is entirely feasible.

b. Seals for Openings. A nuclear reactor plant will have numerous openings in walls for conduit, pipes, and so on. Seals should be of a type that will form a noncombustible fire barrier.

c. Furnishings. Furnishings in rooms adjacent to the reactor should be noncombustible to the maximum extent possible.

d. Protection. Even when all possible precautions are taken, a fire may occur. For such locations the appropriate type of fire protection system should be provided. Because of the potential for severe damage if fire should reach the reactor or its controls, serious consideration must be given to fixed protection even though the risk of fire may be relatively small.

Before leaving the subject of nuclear reactors, one cardinal rule for nuclear safety should be emphasized. *Do not ask the average operating employee what the hazards are or what to do about them.* Questions that fire safety personnel have about radiation hazards at a particular reactor should be asked only of persons who are really qualified to give advice. Such advice should be followed to the letter. (At least two qualified people are on duty at all times in the control room of an active reactor.) If circumstances prevent following through on original advice, new instructions should be obtained from on-duty qualified personnel. The fireground at a reactor is no place to experiment.

In summary, handling a fire emergency at a reactor requires preplanning by both fire safety and radiation personnel. Each should recognize the other's expertise in their respective fields. If they do, the risk of fire can be reduced and the potential for damage in the event of fire held to a minimum.

13

Radiation Machinery

For the purposes of this chapter, the term "radiation machinery" includes all the many types of particle accelerators. We need not distinguish the various types, however, as the basic fire protection problems are the same for all machines.

Most of these radiation machines are used by research scientists for studies in atomic disintegration, transmutation of elements, and so on. Some are used for the production of special radioisotopes. Still others are used for radiography or medical therapy.

In general, particle accelerators are large, complex, and costly. To a large extent, the high cost of these machines is due to their being custom-made to meet particular needs. By their very nature particle accelerators are extremely susceptible to damage by fire, soot, water, excessive heat, and other fire byproducts. Fire safety for this equipment, therefore, starts on the drawing board with careful attention to the fundamentals of fire prevention and protection. It continues with preplanned firefighting procedures in the event fire does occur.

All particle accelerators are designed to impart tremendous speeds to electrically charged particles which are fired at a target. (Figure 55 is a schematic illustration of a linear accelerator.) During operating periods, this process generates considerable radioactivity. During nonoperating periods, there is for all practical purposes no radiation hazard to personnel other than what is being emitted by the target. Responsible fire safety personnel should, therefore, know the target locations and how to shut down the machines without being exposed to radiation.

Basic areas of concern are as follows:

1. The presence of extensive high-energy electrical equipment.

2. The high level of radiation during operating periods.

3. The possible use of large amounts of paraffin or other combustible material for shielding.

4. The presence of large amounts of cooling oils (in the electrical transformers or in the machine proper).

5. The presence of one or more radioactive targets.

6. The possible presence of combustible construction and/or combustible occupancy.

7. The need for preplanning.

8. The possible need for some form of installed fire protection.

9. The possible need for special ventilation.

It is customary to locate particle accelerators in large fenced-in areas where distance provides protection, behind substantial earthwork enclosures, or in concrete

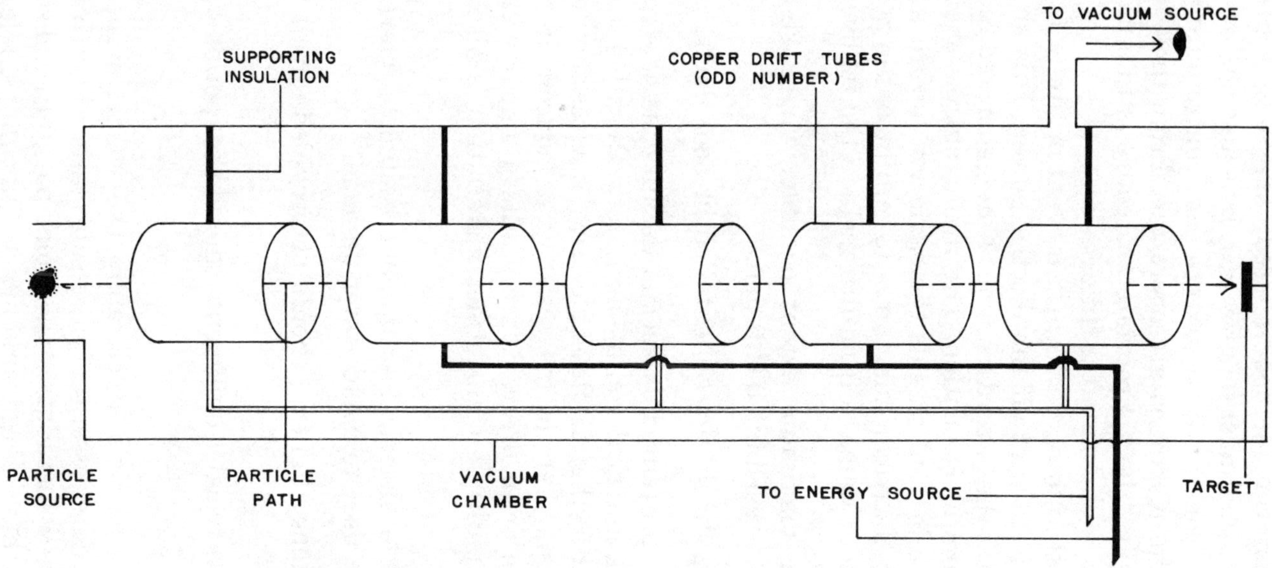

ELECTRONIC LINEAR ACCELERATOR

FIGURE 55

vaults. Whatever method is used, adequate security against intrusion must be provided. It is also customary to require constant attendance at control centers whenever the machines are operating. From these two starting points, the appropriate procedure is to analyze the problem areas one at a time.

The first step to be taken in the event of fire or other emergency is to shut off the electric power. This single act eliminates the hazard from machine-generated radiation and from electrocution. The main controls should, therefore, be located outside the radiation hazard area. Other basic electrical safeguards to minimize the possibility of fire or damage from fire include the following:

1. To the maximum extent possible, wiring should have flame-resistant insulation.

2. Wiring should not be installed in floor ducts.

3. Unless prevented by special design criteria, transformers should be dry or askarel types. Oil-filled transformers and other oil-filled electrical devices should be housed separately in fire resistive or noncombustible structures or in fire resistive vaults. Compartmentation of a multiple transformer installation is also desirable.

4. Valuable electrical equipment should be elevated above the floor to avoid water damage.

5. Emergency lighting should be provided for all key areas.

6. All electrical equipment should be periodically tested.

Most new particle accelerators use water, a special concrete, or other essentially fire-safe material for shielding. A few new installations use some paraffin, and some older units have large amounts of paraffin shielding.

Extreme care must be used in handling firefighting operations when paraffin is present in large amounts. At an early stage in a fire the paraffin will liquefy and give off white vapors with a relatively low autoignition temperature. Accordingly, entry into a vault or room where paraffin vaporization is taking place should be delayed until the vapor flash has occurred or the area cooled down below the vaporization point.

Many particle accelerators are oil-cooled. Quite commonly these units are protected by automatic carbon dioxide systems. Local personnel and responding fire departments should be familiar with both automatic and manual controls for such systems. If personnel work in the protected area, there should also, in the author's opinion, be a predischarge alarm.

As previously noted, even though the accelerator has been deenergized, the atomic target is still radioactive. It is essential, therefore, that the location of the target(s) be known and also the type and intensity of the radiation being emitted.

Unfortunately, some of the earlier machines were installed in buildings that are wholly or in part combustible. Preferably, these buildings should be replaced by fire resistive or noncombustible construction. Often this desirable replacement will not be possible because of overriding technical or economic considerations. For such locations, roofs, ceilings, floors, walls, and partitions should be sheathed with five-eighths-inch or better gypsum board on metal or rock lath or by other materials with equivalent fire resistance. In some cases automatic sprinkler protection may also be needed.

Some occupancies related to the actual operation of the accelerator may be in part combustible and should if possible be located outside the accelerator area. Appropriate fire protection should also be provided for areas with appreciable combustible occupancy.

Many pieces of equipment in accelerators are highly

susceptible to damage from corrosive gases released from such things as burning oils. Special ventilation may be needed to minimize this type of damage. Where such conditions may occur, firefighting forces should be prepared to mount a rapid attack on the fire with water fog. Techniques for restoring wet-down electrical equipment are good enough to assure nearly 100 percent recovery in a short period of time.

Again, the key to handling a fire emergency at a particle accelerator successfully is preplanning. From responsible and knowledgeable officials firefighting forces can obtain essential information on atomic target locations, radiation hazards, locations of main electrical controls, names and addresses of officials to be contacted in an emergency, information about equipment that is particularly susceptible to fire damage and so on. From the firefighting forces, officials at the installation can learn what actions they should take in event of fire, how to eliminate or minimize existing fire hazards, and sources of assistance for restoring operations after a fire. Working together, they can still further improve the already excellent fire record in the field.

14

Nuclear Weapons

Although at first glance the subject of nuclear weapons may seem out of place in a civilian fire safety radiation text, a careful review of the subject will show that it does need consideration. There are thousands of these weapons in place or in transit by land, water, or air. In addition, weapons components are manufactured in many plants and these are of necessity transported to places of assembly. Fire safety personnel, particularly firefighting forces, should, therefore, have a basic understanding of the real radiation and fire hazards.

Far too many people have the idea that if a nuclear weapon is involved in a fire an atomic holocaust will result. This is a totally false assumption that must be eliminated before the actual radiation and fire problems can be dealt with realistically.

Nuclear weapons are designed with extreme care to eliminate the possibility of an atomic explosion. Typically, a nuclear weapon consists of a mass of nuclear material surrounded by high explosives plus specially designed hardware. It is this hardware, acting on the high explosive, that sets up the sequence of events leading to a nuclear explosion. Without this sequence of events there may well be fire and radiation hazards, but there will be no atomic explosion. Designs of control hardware vary to fit the different types of weapons, but all systems are designed to be "fail-safe" in the event of accident. The fact that the design criteria are sound is proven by the record of no inadvertent or unintentional nuclear detonations over a period of nearly thirty years.

Because of the volume of weapons handled every day, the possibility of involvement in fires or accidents exists and such incidents have occurred. The concerns of fire safety personnel are fire and radiation hazards that may develop from such incidents. In fires or accidents, the two weapons components that constitute the most likely and most severe hazards are the high explosives and the nuclear material.

High Explosives and Nuclear Materials

When high explosives are ignited by fire, they will burn and they may explode. Similarly, if a weapon is accidentally dropped from an aircraft, a truck, or any other vehicle, the high explosive may ignite and may explode. These explosions will not make the nuclear material go critical, but they may cause radioactive material to be spread over a considerable area.

Burning high explosives have very definite characteristics with which responsible firefighting forces should be familiar. Principal characteristics are:

1. *Torching*. High explosives, when burning, oxidize at high temperatures and as a result "torch." While some other materials torch, most do not. This, then, is one way to recognize when the high-explosive component of a weapon has become involved in fire.

2. *Smoke.* The smoke from burning high explosives is light in color and is therefore readily distinguishable from the darker smoke from other burning materials in a weapons fire.

3. *After-Fire Hazards.* As high explosives burn, depending on the type, they melt, drip, flow, and spread around. In the process, such unburned high explosives may mix with the surrounding ground or wreckage (quite possibly including radioactive material). After a fire is out the explosives can be considered safe only if they have been completely burned. WARNING: Unburned high explosives remaining after a fire form an extreme explosive hazard. In fact, they may be more dangerous after a fire than they were before. Removal of the unburned high explosives and other debris after a fire should be done only by specially trained explosives disposal and decontamination teams.

4. *Explosions.* The possibility of a detonation is present any time high explosives are involved in an accident. Such detonations may be small or large. The high explosive may be locally scattered in the event of a weapon breakup or a small explosion. These scattered pieces are dangerous and should be removed only by specially trained personnel.

5. *Fire.* Following an accident, a weapon may be intact or may have been broken open. When heated or burning high explosives are confined as they are in an intact weapon, a detonation can occur at any time.

Following an accident or an explosion, nuclear materials become dispersed as finely divided particles. If a fire also ensues, these particles will be dispersed as radioactive oxides. These materials, released from the weapon, are alpha emitters. Basic safeguards against the internal exposure hazard generated by alpha emitters are outlined in Chapter 6, "Internal Radiation."

For firefighting guidance, explosives are divided into four groups, numbered from 1 to 4. The hazard increases with the number. In general, nuclear weapons containing high explosives are in the Symbol 4 group. These are materials that may detonate in event of involvement in a fire. Vehicles transporting Symbol 4 weapons will carry appropriate explosives markings and will also be accompanied by a military courier.

Fire symbols and firefighting practices for explosives were developed long before nuclear weapons existed. Following these practices without modification could result in the unacceptable abandonment of a nuclear weapon that could have been saved by the use of modified firefighting tactics, thereby averting a possible disaster.

Preplanning for Firefighting

Preplanning for the safe handling of nuclear weapons fires is an absolute must. The potential risks to life and property are too great to postpone decision-making on firefighting procedures until personnel are on the fireground. Firm decisions should be made and standard operating procedures developed in advance for handling possible types of weapons fire situations.

Normally, nuclear weapons in transit are accompanied by one or more military couriers who have been trained in actions to be taken prior to the arrival of firefighting forces. However, it is possible for the courier to become incapacitated in an accident and be unable to take action or provide information. It is also possible for a nuclear weapon to be accidentally dropped from an aircraft. Such incidents are rare, but they do happen and cannot be ignored. Therefore, responding firefighting and rescue personnel should include in their standard operating procedures a requirement to notify the nearest military installation or Energy Research and Development Administration regional office.

The Time Factor

For both firefighting and personnel rescue operations, the critical factor is time. How soon will an explosion occur? The amount of time available depends on the makeup of the weapon casing and the intensity and nearness of the exposing fire. It may be only a few minutes or it may be a considerable period. For example, assuming the weapon did not immediately detonate from either impact or the initial fire:

1. When burning high explosives are confined in an intact weapon, detonation can occur at any time.

2. Time becomes extremely critical when a weapon is enveloped in flames or is being subjected to excessive heat by convection, conduction, or radiation.

3. The high explosives of an intact weapon are vulnerable to detonation when the weapon has become so heated it cannot be touched with the bare hand.

4. It is possible for a weapon, intact or broken open, to be near a fire without being engulfed in flames or subject to excessive heat from conduction, convection, or radiation. In such situations, attempts should be made to keep the weapon cool while extinguishing the exposing fire.

Because of the combined radiation and high explosives hazards, judgments regarding the point at which firefighting operations are possible should be on the conservative side. When an explosion is thought to be imminent, all personnel and apparatus should be withdrawn 2,500 feet. (Most references suggest 2,000 feet clearance, but, since flying objects can travel farther than that, the author prefers the extra margin of safety.)

Firefighting Procedures and Equipment

On arrival at a fire in which nuclear weapons are involved, one of the first actions to be taken is clearing the area of all persons not engaged in firefighting or personnel rescue operations. A clear distance from the nuclear weapon of at least 2,000 and preferably 2,500 feet should be maintained. Rescue personnel should stay out of the fire area except for the purpose of saving lives and moving injured people out of the danger zone. In performing rescue operations, all available means should be used to prevent smoke from entering eyes, noses, and throats.

Standard firemen's bunking clothes, boots, helmets, and gloves provide excellent shielding against any alpha radiation that may be present. In addition, a self-contained breathing apparatus and tight-fitting goggles should be worn. All of this equipment may not be available for rescue work personnel, but this should not delay personnel rescue. Reasonably acceptable other means are normally available for the short time rescue work would be underway.

Assuming a decision is made to fight the fire, the senior fire officer present must be prepared to make fast decisions based on the adequacy of the water supply and other extinguishing agents and the manpower available to conduct operations. For example: (1) it may be feasible both to attack the fire and to cool the weapon; (2) exposure of the weapon to heat may be within safe limits and fire extinguishing activities can be concentrated on the fire; and (3) the fire may not pose a direct threat to the weapon or to anything else, but may generate enough heat to require concentration on cooling the weapons.

General procedures to be followed for fires involving nuclear weapons include, but are not limited to, the following:

1. Always approach the fire from upwind and at the maximum angle to armaments (guns, rockets, and

missiles mounted on or in various parts of a wrecked aircraft or other nuclear armed vehicle can be fired by heat from an exposing fire).

2. In making the approach, consideration should be given to:
   (*a*) the maximum efficient use of available extinguishing agents, and
   (*b*) provision for the rapid withdrawal of men, apparatus, and equipment.

3. Programming should provide for cooling the weapon and gaining control of exposure fires as rapidly as possible. (This is one type of fire where possible water damage is of very minor importance.)

4. While water is the most effective agent for cooling weapons and controlling high explosives fires, the method of application is critically important. Water should be applied as a fog or spray. Solid stream application, particularly at high pressures, can be very dangerous, because it can scatter exposed high explosives and radioactive materials.

5. Although foam can be used to extinguish explosives fires, it has some drawbacks. It is not nearly as good an extinguishing agent as water, and it may conceal extinguished, but still dangerous, residues. Such concealment increases the difficulty of cleanup and the possibility that someone may accidentally step on or run over a sensitive explosive. Because of these drawbacks, foam should never be applied to a hot weapon when other agents are available.

6. It should be remembered that high explosives carry all the oxygen they need to burn. For this reason carbon dioxide, dry chemical, and other extinguishing agents that extinguish fires by excluding or displacing oxygen are not effective on high explosives fires. Their

use should therefore be essentially limited to gaining control over fires exposing the weapon.

7. If an explosion is believed to be imminent, firefighting and support personnel should be withdrawn at least 2,000 feet. If possible, mobile equipment should be withdrawn at the same time. Reentry into the fire area should not be attempted until the person in command is certain it is safe to do so.

Following the extinguishing of a fire involving a nuclear weapon, special actions not normally required at fires are necessary. These actions include, but are not necessarily limited to:

1. Withdrawal of all personnel and equipment a minimum distance of 2,000 feet (2,500 feet if possible).

2. Permitting only specially trained explosives disposal and decontamination teams to have access to the cleared area until these persons rule the cleared area to be safe.

3. Gathering all personnel and equipment that may have been contaminated by smoke from a burning weapon into a designated isolation area, located outside the cleared area.

Decontamination of personnel and equipment in the isolation area should be accomplished by specially trained and equipped monitoring and decontamination personnel before release from the isolation area is authorized. This isolation and decontamination program is necessary to prevent possible contamination of others. Additional details on decontamination procedures are provided in Chapter 17, "Radiation Hazards in Firefighting."

In addition to the general procedures outlined in preceding paragraphs, other actions may be necessary in particular situations. Many of these actions could be taken by knowledgeable personnel prior to the arrival of the fire department. Major areas of concern include the following:

1. *Weapons Storage.* Weapons are stored in various types of structures (above and below ground) and outdoors. While tight controls are placed on flammable materials and flame-producing devices, some fire potentials need consideration. For example, in event of fire:

   a. Power from electric circuits should be killed.
   b. Forklift trucks should be removed as appropriate.
   c. Weapons should be removed when possible.
   d. First aid firefighting should be done when possible.

2. *Rail Fires.* Rail fires can occur at any point—at sending or receiving railheads or en route. Derailments in transit can present a considerable problem. Actions that may be necessary include the following:

   a. If at all possible, the affected car should be isolated from other cars.
   b. Interiors of cars on fire should be checked, when accessible, to determine whether or not the interiors have been ignited.
   c. If the interior of a car has been ignited and the weapon is not involved, firefighting with available equipment should be attempted prior to fire department arrival.
   d. Size and weight permitting, weapons not involved in the fire should be removed.
   e. In the event of a derailment followed by fire with the nuclear weapon not being involved, every effort should be made to prevent fire from reaching the weapons car.
   f. In the event of a derailment fire involving a weapons car, evacuation of the surrounding areas should be started immediately.

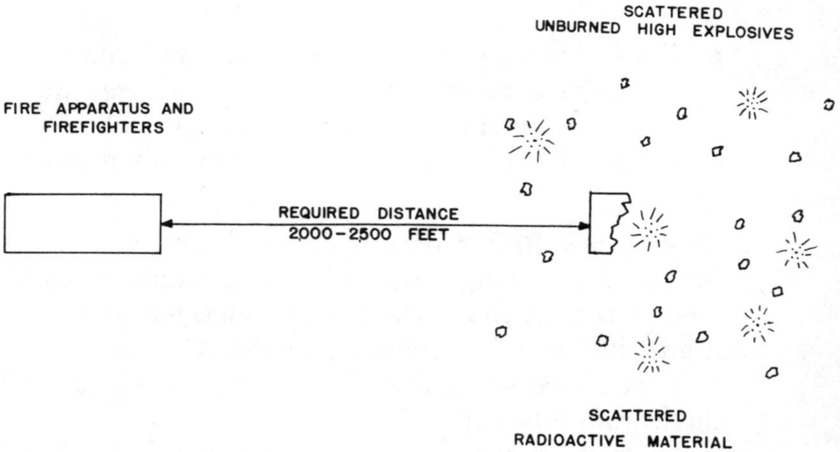

AFTER WEAPONS EXPLOSION

FIGURE 56

g. In all derailment fires, whenever possible, salvage the cargo manifest (the train conductor has this). It can be invaluable to responding firefighting forces in protecting weapons cars as well as other firefighting operations.

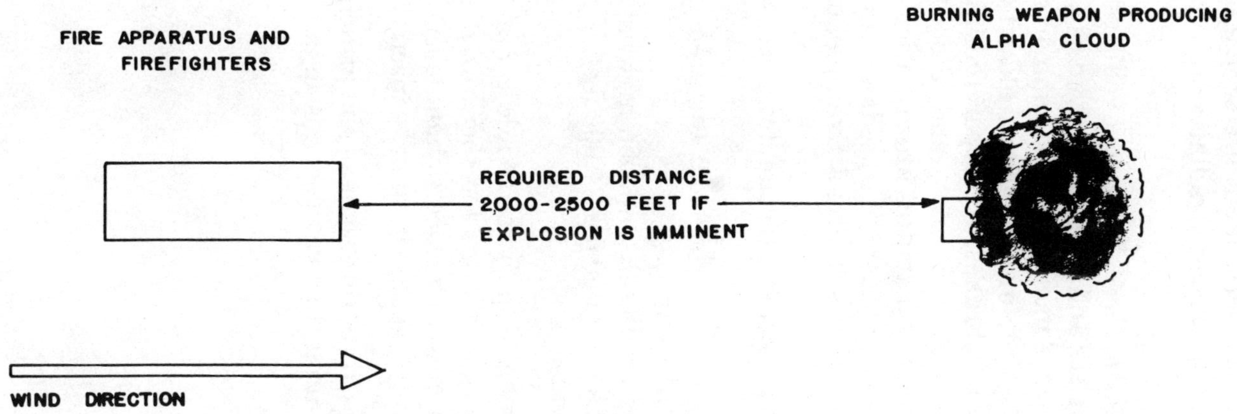
FIRE APPARATUS AND FIREFIGHTERS
REQUIRED DISTANCE 2,000-2,500 FEET IF EXPLOSION IS IMMINENT
BURNING WEAPON PRODUCING ALPHA CLOUD
WIND DIRECTION

FIGURE 57

NOTE. A massive derailment fire or one in a marshaling yard could spread a radiation hazard over a wide area.

3. Transport Vehicle Fires. Transport vehicles come in many types and sizes and may be involved in fires from a number of causes. Among the most common causes of transport vehicle fires are collisions, electrical short circuits, overheated brakes, improper fuelling, overheated tires, broken fuel lines, and the like. The number and variety of potential fire situations is such that courier and driver must have predetermined plans of actions for emergency situations. Examples of actions include:

a. Separating a burning tractor from the weapons van or trailer.

b. Evacuation of personnel from the area when an explosion is imminent.

c. Establishing roadblocks on open highways. Clear distances of at least 2,000 feet (4,000 feet, if possible) should be provided.

4. Training. Firefighting forces should contact appropriate military organizations for specialized training information that may be needed to handle the types of weapons fires likely to occur in their respective areas.

Figures 56 and 57 illustrate some of the principal considerations in handling nuclear weapons fires.

15

Fallout Shelters

While this book was being written, the Chinese conducted a test atomic explosion in the atmosphere. Fallout from this explosion drifted slowly to earth, spreading low-level radiation contamination over large areas. Although the radiation level from this test was too low to constitute any danger, the explosion prompted the author to include a chapter on fallout shelters.

A number of years ago the United States and the Soviet Union, for all practical purposes, were the only nations possessing atom bomb capabilities. At that time both nations conducted atmospheric tests of atom bomb devices. The fallout from these tests plus the "cold war" stimulated a program to determine how to protect the general public from "atomic fallout." Surveys, in which the author participated to some extent, were made, and in many areas fallout shelters stocked with food, water, and medicine were designated. Signs indicating the locations of shelters were put up, and from time to time air raid drills were conducted. As time went on, however, atmospheric tests by the two nations ceased and cold war tensions eased. As a result, the fallout shelter program, except for the protection of highly placed individuals, virtually disappeared.

Today more nations have the capacity to make atomic weapons and still more have access to the "know-how." Many of these nations are not fully aware of the potential for radiation damage from fallout in the event a weapon is used. Although most of these nations can be educated to the hazard so that they will act responsibly, the possibility of an irresponsible group using an atomic weapon cannot be ignored. For this reason, renewed consideration of fallout shelters is, at least in the author's opinion, desirable.

Although the primary purpose of a fallout shelter is protection against radiation hazards, to be effective the shelter must also provide reasonable protection against fire hazards. A shelter that can be made untenable by fire is of no real value in protecting against fallout radiation. In addition, a shelter that provides fire safety for occupants will also provide complete or nearly complete protection against radiation hazards from fallout.

An analysis of fire safety for fallout shelters requires consideration of the following major factors, to be defined shortly:

1. Internal fire exposure.
2. Local fire exposure.
3. Conflagration exposure.
4. Building construction.
5. Fire protection facilities.

Following is an outline of procedures that should be used in evaluating a fallout shelter. Because of the wide variety of situations that might be encountered, this outline is intended as a guide to proper approaches and not as a complete working document.

The first step in evaluating the vulnerability of a proposed fallout shelter is to establish key definitions as follows:

1. Internal exposure is the fire hazard presented by any other occupancy in the shelter building.

2. Local exposure is the fire hazard presented by areas outside of but within 250 feet of the shelter building.

3. Conflagration exposure is the fire hazard presented by areas more than 250 feet away from the shelter building (except when local and conflagration areas hold similar hazards, the entire area is classed as a conflagration area).

4. Construction includes the basic building, horizontal and vertical cutoffs, interior finish, and the like.

5. Protection includes facilities provided to limit the spread of fire.

Because of the presence of radiation hazards as well as fire hazards, protection includes some construction features. Examples of protection features are automatic sprinkler systems, standpipe systems, portable extinguishers, blank walls, smoketight fire doors, and wired glass and glass block windows.

NOTE: A typical plot plan, showing a shelter building and exposure areas, is shown in Figure 58. After a survey of such a situation is made, the sketch would be completed to show essential construction, protection, and exposure details.

Radioactive fallout can be classified into three groups as follows:

1. Severe Fallout. Except for unusually critical situations, firefighting operations by organized fire depart-

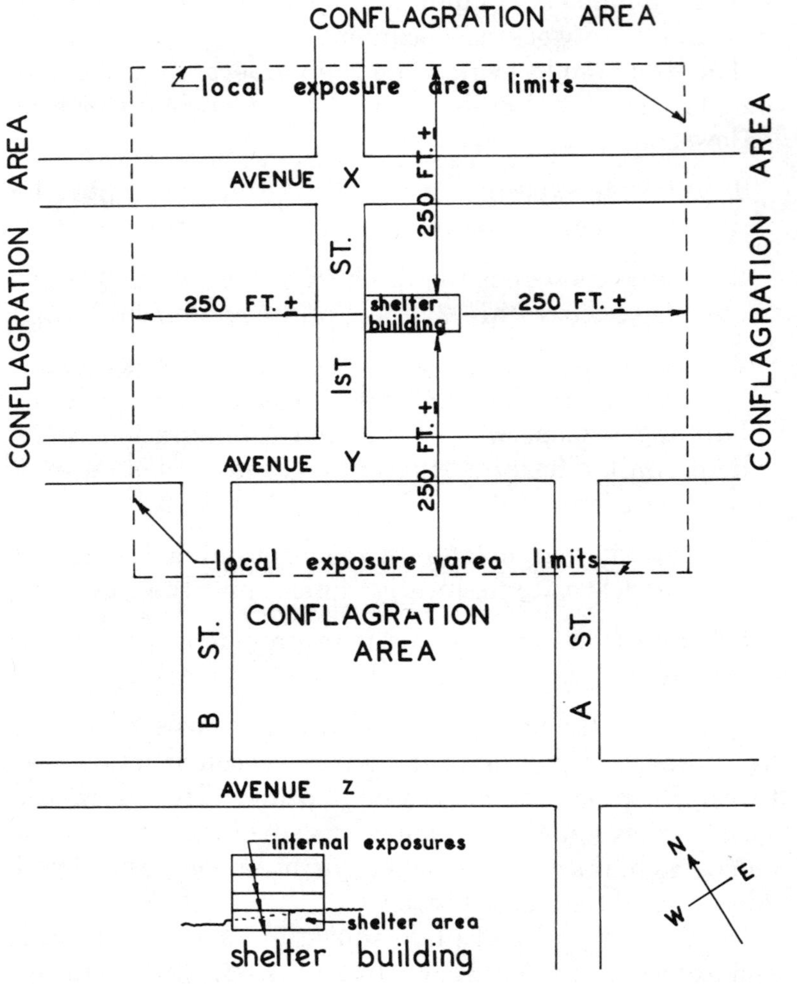

SHELTER BUILDING PLOT PLAN NO SCALE

FIGURE 58

ments are unlikely. Single or multiple fires capable of generating local mass conflagrations without organized fire department response are definitely possible. If major fires occur, it is highly probable that the public water supply will be overtaxed. Therefore, the surveyor should consider that only those automatic sprinkler and standpipe systems that have independent water supplies are reliably in service. Fire pumps, taking suction from open bodies of water that may have been contaminated by the fallout, are also open to question, unless the discharge can be monitored.

2. *Slight to Moderate Fallout.* Some firefighting by organized fire departments would be possible. However, the effectiveness of such operations would depend on the area involved. Single and multiple fires are both possible, but except under unusual circumstances, such as a highly congested area, they should be brought under control before a mass conflagration develops. Public water supplies would not, in most instances, be overtaxed.

3. *No Fallout.* Full fire department assistance would normally be available. The likelihood of public water supplies being overtaxed is remote, assuming the system to be properly designed.

NOTE: Fallout is the radioactive particles dispersed into the atmosphere by an atomic explosion. These particles in dangerous amounts can fall to earth hundreds or even thousands of miles from the site of the explosion. Fallout shelters are intended to provide protection from radiation. They are in no sense to be considered to provide protection against an atomic blast or the ensuing fireball.

Internal Fire Exposure

An analysis of the suitability of a proposed fallout shelter usually starts with a determination of its safety from internal exposures. Determination of the effects of internal exposure is based on building construction details, the character of exposing occupancies, and the adequacy of internal protection systems.

1. Construction ratings are based on fire resistive ratings for roofs, floors, walls, and partitions; adequacy of horizontal and vertical fire cutoffs, flame spread ratings for interior finish, and so on. Such ratings may run from excellent to poor. For example:

a. A proposed shelter building is three stories and a basement and is a Class A fire resistive structure. Fire cutoffs are adequate. Interior finish is noncombustible. Size of fire areas is moderate. Construction rating for a building of this type would be excellent.

b. A proposed shelter building is two stories and an attic of brick, joisted construction. Stairs are open. Interior finish is largely untreated plywood. This building would rate poor.

2. Occupancy hazard classes may range from severe to considerable to moderate to slight to none. National Fire Protection Association Standard No. 13 provides excellent guidelines for rating occupancy. These standards, however, should be modified as appropriate to take care of such items as unusually congested conditions. When a building is of fire resistive construction and fire cutoffs are satisfactory, only the occupancies in stories below the shelter area, on the same story, and on two stories above need to be considered. For other structures it is necessary to consider the occupancies in all stories.

Within the shelter area itself, attention must be paid to occupancy, because in the event of severe fallout, occupants might have to remain there a week or more. Accordingly, shelter occupancy should be designed to be as noncombustible as possible.

3. Shelter building protection includes fire cutoffs as well as automatic sprinkler systems, fire extinguishers, and so on. The reason for including certain construction features, such as fire cutoffs and glass block windows, under protection as well as under con-

struction considerations is to emphasize the fact that they must resist radiation penetration as well as fire. It should not be overlooked that areas above, below, and adjacent may present little or no fire hazard and be vulnerable to fallout.

The final step in evaluating internal exposure is to combine construction, occupancy, and protection ratings into a single overall rating, varying from excellent to poor. If the overall rating is excellent or good to excellent, the fire vulnerability of the shelter insofar as internal exposure is concerned is satisfactory. Lower ratings would indicate the need for some degree of improvement.

Local Fire Exposure

Determination of the effects of local exposure on the proposed shelter requires consideration of three major factors: (1) the burning intensity of the exposure area, (2) the distance between the exposure and the exposed shelter building, and (3) the protection provided against the exposure.

Examples of burning intensity are as follows:

1. Slight. In this category would be noncombustible or fire resistive schools, hospitals, office buildings, light brush, and the like.

2. Moderate. In this group would be:

a. Apartment buildings with noncombustible walls, spaced at least 15 to 20 feet apart, and limited to 20 to 24 dwelling units per building.

b. Schools, office buildings, and so on with noncombustible walls, wood floors, and single floor areas not over 10,000 square feet. Height would be limited to four stories.

c. Flammable liquids tanks, diked and with a combined capacity of 30,000 gallons or less.

d. Lumberyards, well arranged and containing 250,000 board feet or less.

3. Considerable. In this group would come warehouses up to four stories in height, having noncombustible walls and single floor areas not exceeding 10,000 square feet, and wood frame institutions with single floor areas not exceeding 5,000 square feet.

4. Severe. In this category are such exposures as:

a. Multistory wood frame tenements, closely spaced.

b. Warehouses and manufacturing plants with floor areas over 10,000 square feet each and with high-hazard occupancies.

c. Large tank farms or lumberyards.

NOTE: The examples of each type of exposure are guides only and should be modified as necessary to meet local conditions. An example: A wood frame building in a normally damp climate does not begin to present the hazard of a similar building in the desert. Since it is always possible for a sprinkler system to be out of service, exposure ratings should disregard the presence of installed protection in the exposure.

After appropriate burning intensities have been established and the distances between exposures and the shelter building noted, local exposure ratings may be determined by using the method shown in Figure 59.

Protection against exposure may be clear spaces, blank masonry walls, wired glass windows, sprinkler systems, fire doors, and so on. This protection is graded from ample to none. If deficiencies can be economically corrected, this fact should be noted. When protection against exposure can be rated as ample, the shelter is satisfactory insofar as local exposure is concerned.

Conflagration Exposure

Conflagration exposure and its potential effects are determined by the general types of construction and occupancy, the density of the built-up areas or of brush and

FIGURE 59

DETERMINATION OF LOCAL EXPOSURE

EXAMPLE: Assume severe burning intensity 60 feet away.

1. Draw a horizontal line from the 60 foot mark to the severe curve.
2. Drop a line vertically to the rate line.
3. The rate is approximately 4. The local exposure is severe.

Assume the same exposure at 180 feet using the same procedure, the rate is approximately 17. The local exposure is slight.

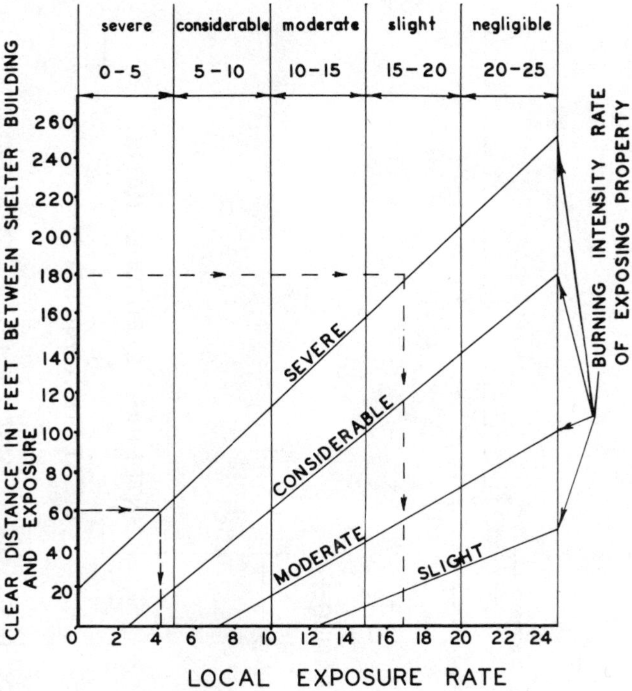

forest areas, the presence or absence of intervening local exposures, and the protection available. (It is perfectly possible for a building to be so designed that it will withstand a conflagration of several days' duration. This has happened many times.) As with other types of exposure, when the protection against a conflagration expo-

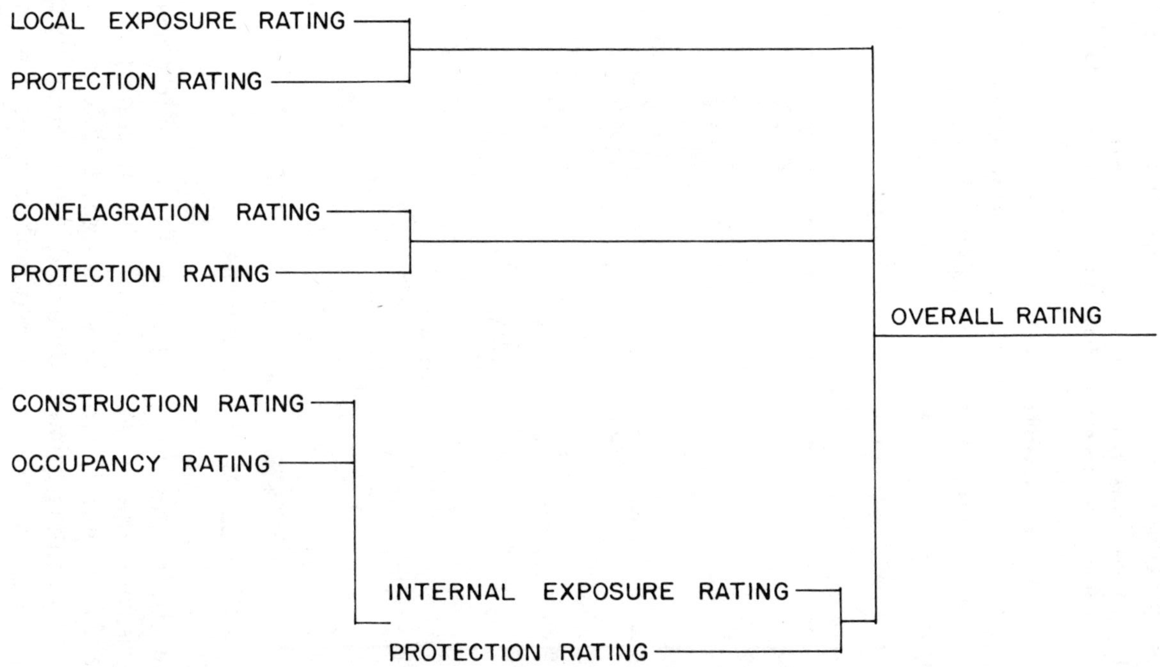

SHELTER BUILDING CONSOLIDATED RATING CHART
FIGURE 60

sure is rated ample, the shelter is satisfactory from a fire vulnerability standpoint insofar as conflagrations are concerned.

All ratings detailed above should be consolidated into a single overall rating which can, if desired, be given a numerical value for record purposes. One way in which the ratings can be consolidated is given in Figure 60.

In pulling all the factors together, it must be remembered that fallout shelters have as their primary purpose the protection of occupants from radiation hazards. The protection period necessary may vary from a few hours to a number of days. During this period occupants need protection from fire as well as radiation. Fire safety features should, therefore, be able to keep both fire and radiation hazards out of the shelter area.

16

Prefire Planning

Because radiation is odorless, tasteless, and colorless, and because all radiation has the power to damage living tissue, prefire planning is a must wherever radioactive materials are or may be present in potentially harmful amounts.

Agency Responsibilities

Responsibility for safe handling of such materials rests jointly on regulatory bodies, users of the materials, transporting agencies, fire departments, and police departments. For example:

1. Regulatory bodies are responsible for establishing and enforcing design standards, setting limits on the amounts of materials permitted to organizations or to individuals, and establishing standards for the use of radioactive materials.

2. Users of the materials are responsible for compliance with regulations, training employees, provid-

ing fire and police departments with essential information, having qualified personnel on duty whenever radioactive materials are present, and maintaining appropriate contact with organizations having radiological assistance capabilities.

3. Transportation agencies are responsible for the provision of appropriate and properly maintained vehicles, the instruction and training of vehicle operators, and cooperation with the public safety authorities having jurisdiction.

4. Fire departments are responsible for:

a. Preplanning with management to minimize the risk of fire and in the event of fire to minimize the hazards of radiation injuries to themselves and others, to limit radiation contamination from fire to the maximum extent possible, and to assist in cleanup operations after the fire.

b. Providing basic radiation training for all firefighting personnel and specialized training for selected individuals. These specially trained individuals should be capable of meeting management on an equal level and jointly working out programs for controlling radiation fire hazards.

c. Coordinating activities, as appropriate, with police departments, public health and safety organizations, and hospitals. In the event of radiation injuries, for example, the ability to inform hospital and/or medical personnel of the nature and extent of radiation injuries quickly and accurately can be invaluable.

5. Police departments are responsible for the control of traffic, for evacuations where needed, and for first-aid assistance.

The Role of the Fire Department

Although all of these groups could be involved in a single fire incident, this is not likely. It is safe to say, however, that the responsible fire department will be involved in every important radiation fire incident within its jurisdiction. It is logical, therefore, to approach the prefire planning problem from the standpoint of fire department operations. In a number of instances the information needed and the actions programmed by the fire department are also applicable to one or more of the other agencies with radiation responsibilties.

Today, because of the tremendous rate of technological change, the yearly flood of new materials, the steady increase of computer-controlled operations, and the increasing size of operations where hazardous materials are used, every fire department should have a fire prevention section and in some instances a completely separate bureau. In the latter situation, the bureau should work closely with or through the fire department. The existence of fire prevention organizations presupposes an inspection program dedicated to locating hazards and devising practical means for eliminating or minimizing them. One such hazard is radioactive materials.

Assuming that a sound fire prevention program exists or has been planned, the first step in providing for the safe handling of potential radiation fire incidents is to select one or more individuals for specialized training and education in radiation problems. The number of such specialists would be governed by the size of the community and the size of the problem.

Fire Department Radiation Officers

The selection of individuals for special training should be based on their educational backgrounds, their ability to absorb scientific data, and their ability to relate to members of the department and the public—not on their initial locations in the department. If more than one person is involved, it is preferable to have both firefighting and fire prevention sections represented. In small departments requiring only one specialist the individual selected should, if possible, have both firefighting and inspection experience.

Training for specialists should begin with instructors' courses offered by agencies of the federal and state governments. Attendance at meetings and conferences where radiation problems are to be discussed should be encouraged. To keep abreast of current thinking and to have a ready source of reference, specialists should be provided with a good small library. Reference materials should include appropriate publications of the Energy Research and Development Administration, the National Fire Protection Association, the Factory Mutual System, and similar organizations. These publications focus on fire safety problems and are not intended to produce nuclear physicists.

In addition to radiation training, specialists need to have the ability, inherent or learned, to deal effectively with people of widely varying backgrounds and responsibilities. This may mean in some instances that attendance at a special management course would be desirable.

Trained specialists have many duties, among which are the following:

1. Representing the fire department in conferences with other agencies to resolve radiation problems of mutual concern.

2. Being the principal training officers in the instruction of firefighting personnel relative to radiation hazards and the proper procedures to be followed before, during, and after a fire.

3. Assisting in the preparation of prefire plans.

4. Acting as the radiation monitoring officer at fires when the presence of dangerous amounts of radioactive materials is known or suspected.

5. Providing essential liaison between the fire department and other emergency services.

After the radiation specialists have been selected and trained and have had time to get a smooth-running program underway, fire departments should publicize their programs and invite the public to make use of their services. In no case, however, should there be publicity other than a simple statement that such activities are being organized until the system is working. Public statements that show the radiation officers to be poorly informed can set a potentially good program back several years or even permanently. It is essential that fire department representatives be able to speak and write effectively from a basis of real knowledge. For questions beyond their experience and competence, these individuals should not hesitate to ask and get advice from qualified personnel prior to commenting on the subject.

Once the need for one or more trained radiation specialists is acknowledged, the next step is to determine how many are needed (for example, one part-time, one full-time, or two or more). To decide, it is necessary to know how widespread the radiation problem is, how much of the firefighting force will require basic training in radiation hazard procedures, and the requirements for plan review. The initial survey may be made in many ways. Perhaps the most common is by fire prevention inspections or surveys. Through these visits the extent and potential severity of hazards can be determined. At this point the fire chief should be able to decide how many man-years will be required to establish and maintain an effective program for the training of firefighters, radiation fire prevention inspection activities, maintaining essential liaison with other concerned parties, and setting up educational programs.

Assessing Hazards in Advance

In protecting against radiation hazards from fire, the first goal should be to minimize the possibility of fire. The second goal should be to prevent small fires from becoming large and to minimize the possibility of radioactive materials being involved in any fire. Recognizing that in spite of all efforts some fires will occur, the final step is to assure that the necessary information and equipment are available to protect responding firefighters from radiation hazards and to limit the spread of radiation contamination during and after fires.

Given the multitude of new materials constantly coming on the market, the wide variations in construction and in protection systems, and the variety of uses to which radioactive materials are put, there is no way in which a fire officer (regardless of experience) can accurately size up radiation hazards on arrival at the fireground without preplanning. Where procedures have been mapped out in advance, however, it is possible to mount an immediate, effective attack with minimum risk to personnel and property.

Preplanning sessions should cover all the basic radiation hazard problem areas plus the actions to be taken in event of fire. Among the most important problem areas are the following:

1. Locations of radioactive materials in the building.
2. Types of radiation emitted by the various sources.
3. Radiation levels: shielded and unshielded.
4. Shielding susceptibility to fire.
5. Warning signs: if signs are combustible or likely to

become involved in a fire, availability of other means to locate the hazards.

6. Radioactive material container susceptibility to damage by fire or heat.

7. Combustibility of radioactive materials.

8. Reactivity of radioactive materials with water or with specific fire extinguishing agents.

9. Likelihood of radioactive material being outside its shield during nonworking hours. If out, availability of:
   a. Means to drop it back within its shield.
   b. Signs to indicate when source is exposed.

10. Possibility of airborne contamination.

11. Availability of holding tanks or other provisions to contain contaminated runoff.

12. Need for contamination holding areas for clothing and equipment used during a fire.

13. Radiation instrumentation needed and its availability. Included are instruments for measuring the various kinds of radiation plus personal dosimeters and/or film badges.

14. Need for self-contained breathing apparatus and/or special clothing.

15. Need for a radiation level check before entering a building.

16. Coordination required with police and health agencies.

17. Availability of lists of qualified persons who can respond quickly with radiological advice and assistance.

18. Last, but not least, establishing a good working

relationship with management. This is of vital importance if potentially troublesome situations are to be satisfactorily resolved to the mutual benefit of both management and firefighting forces. For example:

a. Few management officials have a real appreciation of how fast a fire can spread. Firefighting representatives should be sure to correct this lack of knowledge so that there will be no delay in mounting firefighting operations. Fire department personnel should not be expected to expose themselves unnecessarily to radiation hazards while awaiting the arrival of qualified experts. Neither should the general public be exposed to a rapid fire spread plus possible radiation hazards and emergency evacuations because of unnecessary delay in fireground operations. In short, management needs to be convinced of the necessity of keeping the fire department advised on key points.

b. Fire department personnel should recognize the problems of management and be willing to explore more than one approach to a problem. The final solution should be one that provides for effective firefighting, safety for personnel in the area, and the least possible interruption of the normal activities of the plant and of the general public.

Preplanning for Specific Situations

After there has been a meeting of the minds between fire department and management representatives, the final stage is to preplan for specific situations. In so doing, it should be understood that methods used to train and equip firefighting forces properly will vary with the size of the problem, the size of the city, and the fire department organization. For example:

1. In a small community with one radiation hazard area and a twenty-man fire department, the program would involve the entire department.

2. In a large city multiple radiation hazards of varying types might exist, either concentrated in one district or spread over a wide area. In such situations detailed training of all personnel and equipping all apparatus to meet all types of situations is unnecessary, diverts training time from other needed programs in areas with no radiation hazards, and spreads resources too thin to be effective. For these cities, specific training and equipment should be concentrated on sections of the department most likely to respond to radiation fires.

Preplanning for a specific situation normally starts with the preparation of maps and diagrams plus an orderly rundown of essential facts. This information, together with recommendations for action as prepared by radiation specialists, should be presented to the chief and/or members of the staff for their consideration. These responsible officials should formulate preliminary plans for handling the specific radiation fire emergencies in question.

Too often preliminary plans at the staff level are finalized without considering the potential for worthwhile input from lower echelons of command. This is unfortunate, since frequently such individuals have localized special knowledge about a particular plant area or hazard that is unknown to the staff. Preliminary plans, therefore, should be presented to such individuals for comment and questions. Once this process has been completed and original data and plans revised as necessary, final plans can be developed.

Some fire departments have a tendency to rely on memory for the execution of firefighting plans. Although such reliance may be acceptable at times for such routine occurrences as small brush fires or grocery store fires, it can be disastrous when dealing with a serious radiation fire. Memories may lapse or actions may be taken in the wrong order, and such mental failures are unacceptable

BASIC PRE-FIRE RUN CARD

NAME OF COMPANY ______
ADDRESS OF COMPANY ______
FIRE DEPARTMENT ______

RADIATION FIRE PLAN NO. ______
DATE ______

RADIOACTIVE MATERIAL	LOCATIONS	AMOUNT CURIES	CONTAIN-ERS	TYPE RADIATION	RADIATION LEVEL	HAZARD EXT.	HAZARD INT.

PROCEDURES

REQUIRED EQUIPMENT FOR TRUCK

ADDITIONAL INFORMATION: PROTECTION SYSTEMS, FLAMMABLES, EXITS, ETC.

BUILDING PLAN ON BACK

PERSONS TO BE CONTACTED IN AN EMERGENCY	TELEPHONE NO. DUTY	HOME

FIGURE 61

under the circumstances. Prefire planning, therefore, should be summarized in a written bare-bones outline of all essential data and the proper sequence of action. Helpful, simplified charts and maps should also be made. All this information should be duplicated so that a copy is available for the first-in company and for each responsible chief officer. Sets of information should be in the command cars at all times. Figure 61 illustrates a typical, simple prefire plan.

Following the adoption of a fire plan, the final step is to explain the problem to the firefighters, arranging for their training in actions to be taken in event of fire and providing the reasons for such actions. During this phase, the instructor should encourage questions and be prepared to provide as detailed information as necessary. When there is a serious radiation hazard, the fireground is no place to make a mistake.

In summary, prefire planning is of necessity a cooperative endeavor among fire departments, management, and other interested public agencies. Without such cooperation and an understanding of each other's problems, there will inevitably be occasions when the extent of radiation damage to individuals and property will be far greater than it should be.

No single chapter on prefire planning is complete without a closing cautionary note. The types of potential radiation hazards are far too numerous to be covered in detail in any one chapter or for that matter in any one book. For this reason such chapters should be regarded as basic guidelines on which to build to meet local needs.

17

Radiation Hazards in Firefighting

Although all firefighting activities involve an element of risk, these risks are accepted with confident composure by the professional firefighter provided he has a reasonable knowledge of the hazard and adequate training in the actions needed to protect himself while accomplishing the twin missions of controlling and extinguishing fires plus life-safety rescue operations.

Unlike other hazards that emerged slowly, radiation came into the picture swiftly. As a result, many firefighters are quite understandably hesitant to accept any radiation risk. Their hesitancy is primarily due to a lack of adequate training in protection against radiation hazards. It is the intent of this chapter to remove these quite understandable fears by providing basic principles for the successful and safe handling of radiation fires. With this framework plus a preplanning program, radiation fireground problems can be approached with confidence.

Review of Basic Principles

Before discussion of radiation fireground situations, it is essential that fire service personnel have a firm grip on the following basic facts.

1. Radiation is simply another hazard. It should be neither feared nor ignored. It can and should be understood so that fire radiation problems can be properly and safely handled.

2. To assure proper protection against radiation hazards, preplanning is absolutely essential.

3. Radioactive materials in ordinary use cannot make other materials radioactive.

4. In fire situations, the most serious problem stemming from a radiation incident is usually contamination.

5. Explosions during a fire radiation incident will not be atomic bomb type explosions. They will simply be chemical or dust explosions, identical with explosions in nonradioactive forms of the same materials.

6. Safety measures surrounding the transportation of nuclear weapons preclude the possibility of a nuclear explosion.

7. Radiation control systems for industry, institutions, and other activities require properly qualified people to be on duty in addition to suitable, well-maintained control devices.

8. The distinction between internal and external radiation hazards must be clearly understood.

9. Radiation has power to destroy or damage living tissue.

10. Federal government records conclusively prove

that it is possible to handle radioactive materials on a large scale without seriously endangering personnel.

11. Acceptable radiation risks are those in which the benefit to be gained outweighs the cost of providing essential protection.

12. Within known limits the body has the capability to repair radiation damage.

13. Radiation sickness results from a massive overdose of penetrating, long-range external radiation. Such overdoses may result in death, permanent injury, or long illness.

14. Radiation injury results from exposure to smaller doses of penetrating external radiation or from exposure to less penetrating radiation.

15. Radioactive poisoning results from dangerous amounts of certain radioactive materials lodging in the body.

16. Radiation damage to the body from long-range, penetrating radiation results when the radiation strikes an atom and knocks an electron out of orbit.

17. Roentgens are used to measure penetrating external radiation in the air. One roentgen of exposure equals one rem of effect on the body.

18. Industrial workers are permitted an annual exposure equivalent up to 5 rem in effect per year. Fire service personnel should average a dosage of well under 5 rem per year. However, a one-time-only dosage of 25 rem is acceptable in an emergency situation.

19. Protection against external radiation hazards can be provided by:

a. Time. Radiation levels are usually given in number of milliroentgens per hour. The dosage effect from a given exposure will vary directly with the time exposed.

b. Distance. The effect of radiation varies inversely with the square of the distance from the source.

c. Shielding. Dense materials placed between the source and personnel stop some radiation. The thickness of the material that will stop half the radiation is called the half-value layer. Each succeeding half-value layer will stop half the remaining radiation. To be effective, shielding integrity must be maintained. Cracked, damaged, or sagging shields lose much of their value.

20. Definitions of interest to firefighters include the following:

a. The *curie* is used to measure the rate at which a material is disintegrating. It measures all types of radiation. Since radioactive materials do not all emit the same type of radiation, it is necessary to know the material as well as the number of curies.

b. Alpha particles are composed of two protons and two neutrons each and carry a positive charge. They lose their energy in about 3 inches of air and cannot penetrate the human skin. Hazard is from ingestion, inhalation, entry through open cuts, or possibly through skin absorption.

c. Beta particles are free electrons carrying a negative charge. They can pass through about 100 feet of air and up to 1 inch of denser materials. Few beta particles can penetrate more than one-half inch of body tissue. External hazard is primarily from exposure of a large body area to a strong beta source.

d. Gamma radiation is essentially pure energy and can penetrate several hundred feet of air and up to 3 feet of high-density concrete.

21. Radioactive materials may emit only one type of

radiation or they may emit more than one type. This is a major reason for identifying the material.

22. External radiation hazards are fairly easy to control. Accurate means for measuring the level of radiation and positive ways to protect against the hazard are available.

23. Controlling or eliminating the possibility of radioactive material becoming waterborne or airborne goes a long way toward minimizing the risk of internal exposure. Some of the principal means for hazard control are:

a. Good building construction (fire resistive or noncombustible rather than wood frame).
b. Sound design and controls for ventilating and air-conditioning systems (noncombustible materials, means to prevent radioactive particles from leaving the building, access to controls for fire service).
c. Good process enclosure construction, including adequate separation from other areas.
d. Strippable wall films and bench covers.
e. Tight limits on the amount of combustibles permitted.
f. Holding tanks for sprinkler and/or hose water runoff.
g. Good enforced rules against smoking, eating, and drinking in hazard areas.
h. Provision for special clothing where needed.
i. Available breathing apparatus of the proper type.

NOTE: These items and others as appropriate should be a matter of concern in fire prevention inspections and plan review, in prefire planning, and in training programs for firefighters.

24. Radioactive materials have half-lives varying from a few minutes to thousands of years. The half-life

is the time required for a material to have its radiation emission level cut in half.

25. Film badges and dosimeters, used to measure radiation exposure, do not measure alpha radiation. Where alpha radiation is present, special measuring devices are needed.

26. Internal radiation exposure problems depend to a large extent on the half-life of the material and whether or not it is soluble in the digestive system.

27. Of all the readily available radioactive materials only uranium 235 and plutonium are fissionable, meaning that they can participate in a chain reaction.

28. Sustained chain reactions do not occur under natural conditions. They are produced under controlled conditions such as are produced in a nuclear reactor.

29. Sustained chain reactions as found in industry will not produce an atomic bomb type explosion. For that to occur, special hardware that is not present is required.

30. For the fire service in general, factors of primary concern regarding nuclear reactors are:

a. The type of moderator or reflector used. Loss of water while running at full power would result in severe damage from residual heat plus possible widespread contamination.

b. The heat exchange medium, which may be water, liquid sodium, or some other combustible metal. Since sodium and some other combustible metals will burst into flame on contact with water, the presence or absence of safety dump tanks and/or warning signs is a matter of concern.

c. Safety of controls for shutting down a reactor. The availability and accessibility of standby controls is essential so that the reactor can be controlled if fire or other emergency prevents the use of primary controls.

31. Properly designed particle accelerators will have a main electrical disconnect switch, located outside the artificially produced radiation field. Pulling this switch will instantly stop the radiation process, leaving only the location of the radioactive target in question. Since large amounts of paraffin are frequently used at particle accelerators, prefire planning should provide for the possibility of a fire of this type.

32. Although radioactive materials in transit, when of importance from a radiation standpoint, are packaged in special containers designed to resist most accidents, the possibility of a package splitting open or being so damaged by fire that there is direct radiation exposure cannot be ignored.

It is of course possible to add to the preceding list. For most fire service personnel, however, these points are adequate as a base for: (1) fire prevention inspections, (2) prefire planning, and (3) setting up firefighting training programs. It should be obvious that close coordination among these three functions is essential to the proper handling of radiation hazards at the fireground with minimum possible risk to personnel and the public and minimum damage to property.

A Hypothetical Case

Based on a firm grasp of the principles behind the safe handling of radiation hazards and a good training program for firefighters, planning for a possible fireground radiation incident should begin. Results of the planning session should be promptly incorporated into the firefighter training program. Failure to do so can easily result in unnecessary exposure of personnel to dangerous levels of radiation and excessive property damage. The importance of tying all the elements of the fire service together is illustrated in the following paragraphs, in which a theoretical radiation fireground hazard is analyzed.

The CBZ Company owns two five-story buildings of brick, joisted construction. Building story heights are approximately 10 feet per story and the buildings are separated by a 21-foot alley. Building *A* has a basement and is 40 feet deep. All stories of both buildings have windows. Building *B* has a sprinkler system, but Building *A* does not. In the fifth story of Building *A* is a radioactive source consisting of 1,000 curies of cobalt 60 in a lead-lined steel container. In addition to the radiation source, the fifth story contains miscellaneous combustible storage. All other stories in both buildings are occupied for metalworking.

Unnoticed when the plant closed for the day, a benzene drum had started to leak. Flammable vapors began to accumulate and some time after midnight were ignited by a pilot light on a gas-fired water heater. About 2:00 A.M. a patron of Smith's Bar and Grill, located a block away, saw flames coming out the fifth story windows on two sides of the building. He called the fire department and strolled over to watch the excitement. The temperature at this time was 50°F., the humidity 40 percent, and the wind velocity 5 mph out of the south.

The city fire department responded to the alarm. If the chief assumed that no radioactive materials were present, he might locate the aerial ladder truck and manpower about as shown in Figure 62. (Pumpers have been omitted to simplify the problem.)

On arrival the fire chief is notified of the 1,000-curie cobalt 60 source. At once he begins to review the normal plans for placing men and equipment from the standpoint of the potential radiation hazard. (By rights this should have been worked out in the preplanning sessions.) To be on the safe side the chief assumes that the shielding for

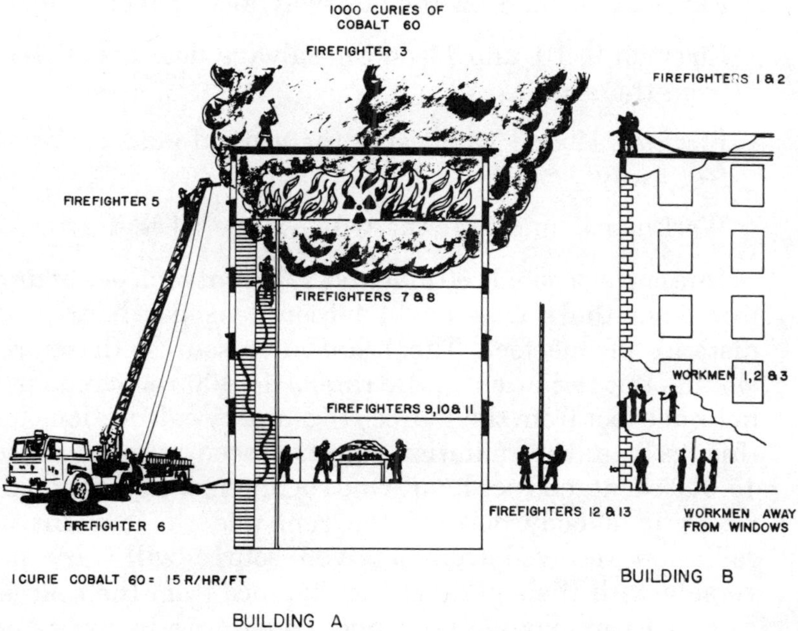

FIGURE 62

the source has failed. How safe are the firefighters? How safe are the workers?

Firemen 1 and 2 on the roof of Building *B* are 45 feet from the source.

Fireman 3 on the roof of Building *A* is 20 feet from the source.

Fireman 4 is at the unillustrated pumper.

Fireman 5 on the aerial ladder is 27 feet away.

Fireman 6 at the base of the aerial ladder is 52 feet away.

Firemen 7 and 8 on the stairway are 17 feet away.

Firemen 9, 10, and 11 on the salvage floor are 37 feet from the source.

Firemen 12 and 13, raising the ground ladder, are 54 feet away.

Workmen 1 and 2 at the window are 54 feet away.

One curie of cobalt 60 gives off 1.6 roentgens per hour 3 feet from the source or 14.4 roentgens per hour at a distance of one foot. The 1,000-curie source, therefore, would emit radiation at the rate of 14,400 roentgens per hour one foot from the source. To simplify calculations for this problem the radiation rate has been rounded off to 15,000 roentgens per hour. One roentgen of exposure will result in a body dose of one rem and the amount of radiation received from a given source will vary inversely with the square of the distance from the source. In this case, exposed personnel remaining in the same place for one hour would receive the radiation dosages shown in Table 8.

Table 8

Personnel	Equation	Exposure Rate	Dosage in Rems
F1 & F2	15,000/45 × 45	7.5r per hour	7.5
F3	15,000/20 × 20	38r per hour	38.0
F5	15,000/27 × 27	21r per hour	21.0
F6	15,000/52 × 52	6r per hour	6.0
F7 & F8	15,000/17 × 17	52r per hour	52.0
F9, F10, & F11	15,000/37 × 37	11r per hour	11.0
F12 & F13	15,000/54 × 54	5r per hour	5.0
W1 & W2	15,000/54 × 54	5r per hour	5.0

Obviously, exposure rates are far too high to be accepted. At this point the chief must make a decision. One answer could be to evacuate the area and let the building burn, as there would be little likelihood of airborne radiation from a cobalt 60 source. If this is done, the chief is assuming the risk of the fire spreading to other buildings. This decision could also seriously complicate the problem of recovering the radiation source after the fire. The next question is whether a way can be found to confine the fire to the building of origin. Proper preplanning would enable the chief to explore all possibilities and to come up with some practical answers. It is possible to control the theoretical fire without seriously endangering the firefighters and with every prospect of keeping down property losses. (It should also be noted that if the chief does not have advance information of the existence of a serious radiation hazard, the potential for serious exposures to personnel is very high. The chief must have this information.)

Assuming the existence of a major radiation hazard is known, the fire service should have qualified people, equipped with high-level monitoring instruments, within the department or readily available from other sources. Response by such personnel may be simultaneous with that of the fire department, or it may be delayed.

Assuming the monitoring service to be delayed, some actions still can be taken to confine the fire to the building of origin. In order to be safe, planning must be done on the assumption that the shielding has failed and that the source is exposed. (It is not safe to assume any important shielding effect from the board floors.) Time and distance are therefore the only reliable allies. Using time and distance wisely, firefighting operations such as the following could be initiated prior to the arrival of complete monitoring equipment.

1. Even under the worst conditions an aerial ladder

company should be able to place a deck gun in service in about 10 minutes (1/6 of an hour). It is known that the radiation level on the ground near the ladder is 6 roentgens per hour. Therefore, the radiation exposure to the ground crew during the setup would be 6 × 1/6 or 1 roentgen. This is an acceptable exposure and would result in a dosage of only 1 rem or 1/5 of the allowable dosage for a year.

2. One man, held in reserve, could climb the ladder to adjust the deck gun and descend within 5 minutes. His exposure, using the maximum radiation level rate of 21 roentgens, is 21 × 1/12 or 1.75 rems, which is well within acceptable limits.

3. As necessary, other unmanned, fixed heavy hose streams could be placed in service.

4. Operations such as salvage, roof ventilation, and manning hose streams inside the building or on the roofs of nearby buildings should not be attempted, pending the arrival of monitoring equipment. These activities would not accomplish enough good to warrant exposing personnel to potentially high radiation levels (for example, unmanned lines from the roof of Building *B* would be acceptable, whereas manned lines would not be).

5. All employees in Building *B* and any remaining in Building *A* should be evacuated.

6. The fire area should be cordoned off and all nonfirefighting personnel, including newspaper and TV reporters and cameramen, kept out.

7. Careful records should be kept of who did what so that, if it became necessary to shift hose streams, no individual would be subjected to an excessive amount of radiation.

Assuming that the proper radiation monitoring equipment is on hand, situations such as the following may be encountered.

1. Building *B* is sprinklered. Near the sprinkler siamese the radiation level with the source fully exposed is 5 roentgens per hour. Monitoring shows an actual radiation level of one roentgen per hour, indicating that 80 percent of the shielding is in place. As long as the radiation level stays at that point, no danger to personnel would occur from manning hose streams for an hour or two.

2. Since it is not known what part of the shielding is in place, the reading at the siamese should never be taken to mean that it is safe to man hose lines elsewhere on the street or inside the building. It is quite possible for the true shielding to be completely gone and for the low reading to be due to intervening accidental shielding such as a storage vault. It is also true that 80 percent shield integrity in one direction is no guarantee that there will be 80 percent shielding in other directions. No prolonged inside or outside operations should be carried on except when personnel can be accompanied by monitors and monitoring indicates it is safe.

3. If monitoring indicates that inside operations can be carried out, there must be two-way portable radio communication. Fire developments in other areas may indicate that loss of the shield or the collapse of the floor supporting the shield is imminent. It is essential that personnel be able to evacuate immediately when radiation levels are likely to change.

4. Continuous monitoring is essential in every area where personnel are working. A radiation reading at one point 20 feet from the source is no guarantee that the level is the same at a distance of 20 feet in any other direction, horizontally or vertically.

5. Because shielding can fail at any time, all persons not actually required for firefighting and/or emergency operations should be kept out of the area. Even though the potential dosage is well within acceptable limits, unnecessary exposure to radiation is poor management.

A second phase of preplanning requires the programming of actions to be taken after a fire has been controlled or extinguished. Failure to do so could well result in serious, unnecessary exposure of life and property to radiation hazards. For example:

1. In the worst conceivable situation the building collapses and the unshielded radiation source is buried in the debris. Under these conditions the only overhaul that can be accomplished is flooding from a distance. Source recovery must be left to qualified experts.

2. In the second worst possible situation the supporting floor collapses, dropping the unshielded source to a lower story. Here again true overhaul, particularly from inside the building, is not practicable.

3. A third possibility would be for the radiation source to become unshielded while remaining in the same position. Again, radiation levels would be too high to permit normal overhaul.

Under no circumstances should the fire department attempt to recover the radiation source after a fire. For this task, assistance must be requested from qualified personnel such as representatives of the Energy Research and Development Administration or the Department of Defense. Locations of a number of offices from which assistance can be obtained are provided in Chapter 9. Pending arrival of assistance, a watch must be maintained to guard against a flareup and to keep the curious away.

In preplanning and on the fireground it is vitally important to consider the possibility of having a critical life-rescue emergency situation. Assume for the moment that the night watchman discovered the fire and collapsed while phoning in the alarm. Assume further that the shield for the radiation source failed simultaneously with the arrival of the fire department and that the radiation level reaching the watchman is 150 roentgens per hour. At most there is only one hour to get him out. Here the chief has a hard decision to make. Thirty men are available for rescue work, but only three men can work in the area at a time. The work involved is only moving debris to reach the watchman, but it will require three man hours. How should the fire chief do it?

1. The 30 firefighters could be divided into ten 3-man teams, each team being given a 6-minute tour of duty. This means exposure of 150 × 1/10 or 15 roentgens per man. If there had been no earlier major exposures, this degree of exposure for life-rescue purposes could be classed as acceptable. It could also be a very inefficient way of doing the job (loss of time in changing shifts plus quite possibly making the entire company unavailable for other radiation emergencies).

2. Eight 3-man teams could be used. Here each individual would work 7½ minutes or ⅛ hour in the radiation zone. Each of these workers would receive 150/8 or 18.75 roentgens of exposure. Six of the firemen would receive no exposure. If this route is selected, who is excluded from the working parties?

   a. The youngest members who are most likely to sire families.

   b. Older individuals with valuable technical knowledge.

   c. Those, regardless of age, who have had major previous radiation exposures.

A fire chief could, of course, find other possible solutions to this problem. The point to remember, however, is that at the fireground the senior fire officer is responsible for conducting rescue operations and for determining how much radiation exposure can be accepted. This officer should always follow a basic principle: Limit the exposure of personnel in all cases to the minimum time necessary to accomplish the task and to involve the minimum possible number of people so that the accumulated units of exposure will be the absolute minimum for the department and no individual will be exposed beyond acceptable limits.

The final stage of preplanning is to take whatever steps are possible to reduce the hazard. With the cooperation of management, these steps might include:

1. Full or partial sprinkler protection for the building housing the radiation source.

2. Relocating the source to a safer location.

3. Providing a fire resistive vault for the source in its present location.

Responding firefighting forces need to know the actual status of any major changes proposed or underway. It makes a big difference to them where the radiation source is located at the moment and under what conditions it is being stored or used. Such knowledge helps the fire department do the following:

1. Determine the proper locations of responding men and equipment so that apparatus will be as far as possible from the actual position of the source and still function efficiently.

2. Establish personnel assignments so that full advantage may be taken of any physical barriers that could be located between firefighters and the actual source location. There are no practical portable shields.

3. Know whether or not plant employees have standing instructions to drop the source back in its shield in the event of fire. If this was not done, arrangements should be made to notify the fire department immediately upon their arrival.

4. Emphasize the value of time. A source that gives off 20 roentgens in an hour emits only 4 roentgens in 12 minutes.

Certain special shielding precautions require mention here:

1. Never disrupt a shield to get at a fire.

2. Even in laboratory hood fires, do not go behind lead brick shielding to extinguish a fire until qualified persons have ruled it safe to do so.

3. When it is known or suspected that radiation sources may be present, view all molten or softened lead with suspicion.

4. Make certain that cracked or spalled concrete walls around a radiation area are monitored.

5. Where vaults have been provided for large radiation sources, learn how to drop such sources into the vault before going into the area. Figure 63 illustrates a typical arrangement. When the barrier is found across the corridor leading to the radiation area the source is out of the vault. Dropping it back into the vault before the area is entered is a must.

The hypothetical example provided in the preceding paragraphs is a very severe case and the likelihood of the average fire department having to face such conditions is remote. Its purpose is to illustrate a number of the key factors that could be present in a radiation fire incident. While very large sources are unlikely in multistory, joisted buildings, sources of this size do exist and are used

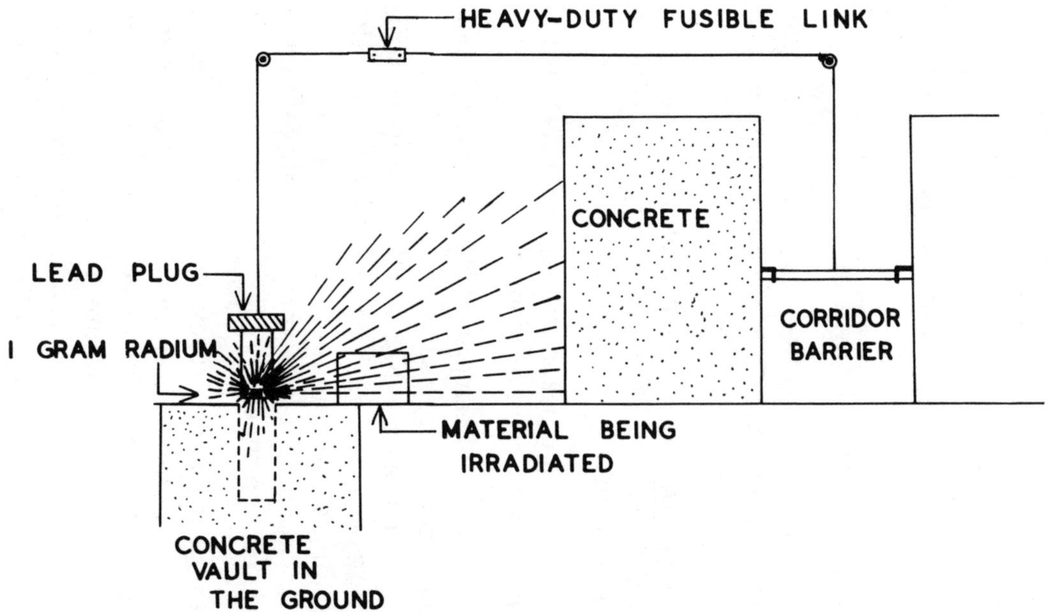

IRRADIATION VAULT AND TYPICAL SAFETY ARRANGEMENT

FIGURE 63

at special locations. They might, therefore, be found in transit or temporarily in transit sheds.

From a practical standpoint, locating a 1,000-curie radiation source in an upper story of a joisted building would require such massive shielding that the floor would probably not support it without prohibitively expensive reinforcement. Within ordinary types of buildings, including occasionally joisted structures, radiation sources of 100 curies or less will sometimes be found. While these smaller sources allow more flexibility in operations in the event of fire, they still represent a potential hazard to life and property. It is, therefore, just as important in such situations to take time, distance, and shielding into consideration in planning and carrying out fireground operations.

Before taking up the question of contamination during and after a fire, a final point on the use of the inverse square law needs to be mentioned. The law assumes that, as is usually the case, only one source is present or that the several sources are essentially at one point. In cases of multiple sources of gamma radiation or wide dispersal such as would occur during a fallout, the inverse square law begins to apply only when the distances from the source area become relatively large. In the few instances where such situations may exist, safety in operations requires multiple monitoring locations.

Handling Contamination

Most radiation problems that the fire service will encounter will not involve large amounts of radioactive materials. In general, these materials will be so located and protected that there will not be a major contamination problem. In a few areas (nuclear reactors, nuclear weapons, and transportation) the possibility of heavy contamination exists unless the fires are handled properly. Responsible fire services should analyze these situations very carefully. Even at lower radiation levels,

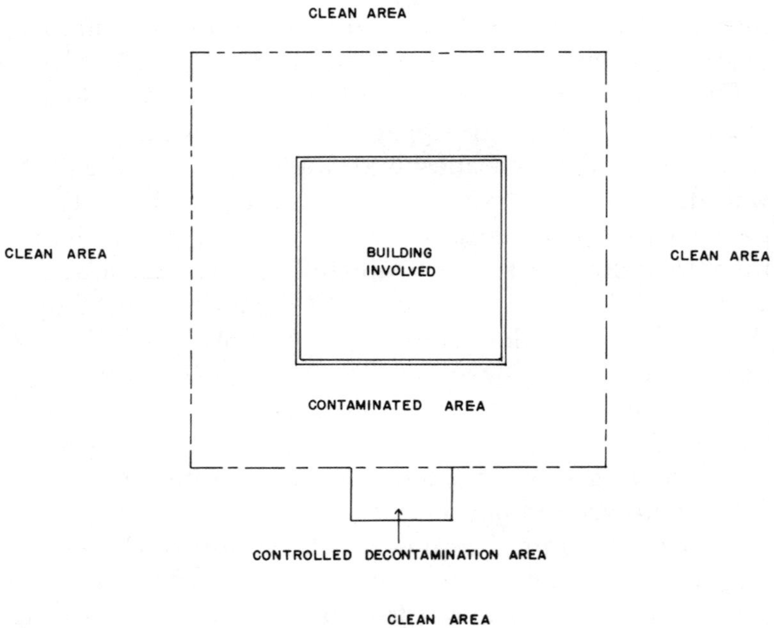

FIGURE 64

a potential contamination problem exists wherever radioactive materials can become airborne or waterborne during a fire. It is essential that responsible individuals be drilled in safe practices so that when a real risk develops, the right actions will be instinctively taken.

There is no sure way of preventing contamination of clothing or equipment used by personnel entering an area where there is a radiation contamination hazard. All such clothing and equipment must be monitored and decontaminated before being returned to a clear area. Unless this is done, there is serious risk of carrying contaminated clothing and equipment back to the station and the contamination to the community at large. Decontamination is time-consuming and may require equipment to be out of service for some time. Accordingly, three measures must be placed in effect immediately upon arrival at the fireground:

1. Establish the line between contaminated or potentially contaminated areas and clear areas and secure this line.

2. Keep the number of personnel and the amount of equipment entering the contaminated area to the minimum necessary to do the job.

3. Personnel and equipment entering the contaminated area must be monitored and decontaminated before reentering the clear area. For personnel this may well mean leaving clothing behind for later decontamination, crossing the line, and donning fresh clothing. It may also mean a washdown before dressing in the new clothing.

While most equipment (hose, ladders, hand tools, breathing apparatus) can be decontaminated fairly easily, this is not true of motorized fire apparatus. Unless it is absolutely necessary, major pieces of fire apparatus should never be brought into the contaminated area. Once contaminated, it could be months before the unit can be placed back in service. This same reasoning applies to any other complex piece of equipment.

Aside from the obvious need to develop safe practices regardless of the radiation level, it is necessary to consider public relations. Even though the radiation level may be well within acceptable limits, the fact that the public knows that this mysterious radiation process is involved in the fire makes it essential that the public be convinced that the fire service (1) is preventing the radioactive materials from getting out into the community and (2) is competent to direct them in the actions they should take for their own protection (evacuation, remaining indoors, and so on).

Under normal fire conditions, fire service personnel with their equipment would move from the apparatus to the fire area. From there it is quite possible that they would fan out to stories above and below the fire, to adjacent areas, and to other fireground positions favorable for fire attack. For various reasons personnel might be moving in and out of the fire area and switching from one fire attack position to another. Backup equipment and hose lines might well be placed close to the fire. If, however, the fire involved radioactive materials, the normal freedom of movement of firefighters could very easily result in contamination being carried into otherwise clean areas. Inevitably, people in the clean area and the public at large would be unnecessarily exposed. In all probability, there would also be serious and unnecessary contamination of apparatus and equipment.

No two fire service organizations will face the same radiation problems. Neither can any two organizations operate in exactly the same way because of differences in personnel, equipment, supplies, climate, and so on. All, however, would or should be using approximately the same approach to radiation firefighting problems. Among the most important procedures and precautions are the following:

1. Park fire apparatus upwind or out of the smoke area. In an area of constantly shifting winds the apparatus should be parked behind an intervening building, if possible.

2. Establish a definite contamination control line beyond which only personnel and equipment that are essential to the fire attack would be permitted.

3. Limit travel in the contamination control zone as much as possible.

4. Removal of needed tools, hose, and other equipment from the apparatus should be done only by persons who have not entered the contamination control zone. These materials should be passed, not carried, across the control line.

5. Radiation exposure records of personnel in the contamination zone should be readily available in the clear zone, so that individuals in the contamination zone can be promptly withdrawn or moved to a less exposed area as necessary. No one assigned to the contaminated area should be allowed to become unduly exposed.

6. All tools and equipment taken into the contaminated area, whether used on the fire or not, should be taken to a designated checkpoint inside the controlled area but close to the security line for monitoring and decontamination before being returned to the apparatus.

7. Hose used in the control zone should be laid out to dry at a designated point inside the zone. After the hose has been dried, it should be monitored and decontaminated before being moved into the clear area.

8. Personnel in the contaminated area must be monitored and decontaminated before leaving the zone. Where clothing contamination is such that it cannot be promptly removed, the clothing must be removed and left behind for later decontamination. Fresh clothing should be available in the clear zone. Prior to crossing over into the clear zone and donning new clothing, exposed individuals must be remonitored. In some instances a washdown to remove residual contamination may be necessary prior to dressing. This is no place for false modesty.

9. Never remove breathing apparatus in a contamination zone until the area has been ruled safe. Replacement of air and oxygen should be done at the control line with backs to the clear zone.

10. If by any chance an individual in the area has breathed radioactive smoke, make sure that person

gets prompt medical attention. If known, advise the medical personnel of the nature of the radioactive material.

11. Where feasible, use water spray instead of solid streams in order to minimize water runoff contamination.

12. Block water runoff to waterways, storm sewers, and so on to the maximum extent possible. If contaminated water does get into water or sewer systems or into open bodies of water, notify the appropriate authorities immediately.

13. In laboratories be especially careful with solid streams, even small ones. Breakage of glass containers for minute amounts of radioactive materials could make the laboratory inoperative for months even though there is no risk to personnel.

14. On outside transportation fires involving radioactive materials, certain procedures as follow should be standard practice.

a. Fight fires from upwind at maximum range.

b. Use ditches or any other available barriers as protection against possible explosion. Even a minor puff that of itself poses no threat can move some types of radioactive particles quite a distance.

c. In most instances, assuming a moderate to large shipment, trained personnel will travel with the shipment. These individuals will have full information on the cargo and normally will also have appropriate instrumentation.

d. If there is a placard on the truck or rail car, immediately call the indicated office. If the placard has been destroyed or is illegible, contact the trucking company or the railroad to determine the nature of the shipment. Simultaneously, the regional radiological assistance office should be called for advice.

e. Request police assistance to secure the area.

f. Initially use solid streams from far enough back so that the stream breaks into droplets. If this does not cause a violent reaction continue this type of application to extinguishment. If reactions occur, discontinue application and allow the vehicle fire to burn itself out. Otherwise protect exposed property to the extent possible.

g. Do no closeup overhauling until the arrival of the appropriate authority and receipt of clearance from this authority.

h. Make certain that guard service is provided to prevent removal of wreckage until it has been cleared for removal by competent authority.

Figure 65 illustrates a sound approach to a radiation fire.

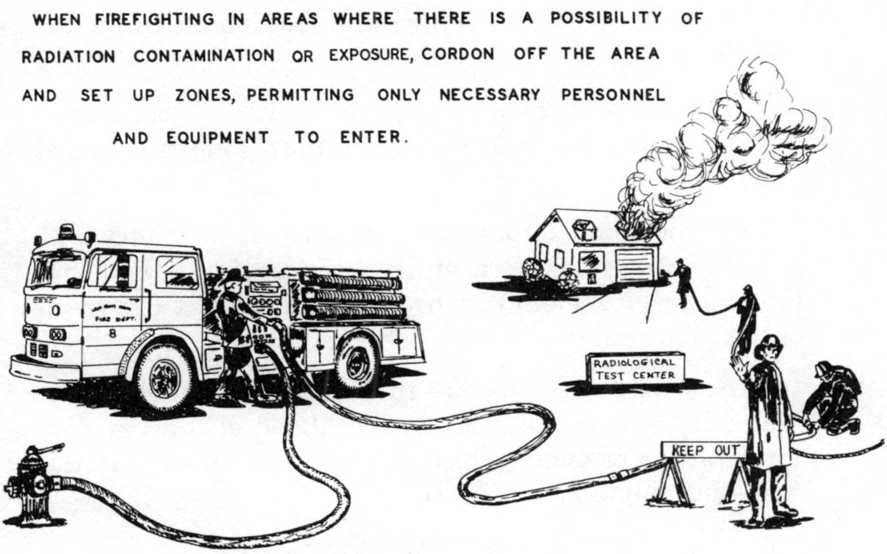

FIGURE 65

Prefire Planning

The firefighting procedures outlined in this chapter are not intended to be all-inclusive but rather to provide a basic framework on which to build a local program tailored to meet local needs. Procedures established should be mandatory. It must be remembered that a radiation level low enough to cause little or no damage during an hour or two of firefighting operations may become lethal or highly dangerous to a person unknowingly exposed to it over a period of days or weeks. Radiation cannot be neutralized.

It must be reemphasized that removal of radioactive contamination can be very costly. For example, careless firefighting procedures might make it necessary to:

1. Junk a building and its equipment that could otherwise have been restored to useful service.

2. Secure a building or area and make it off-limits for days, weeks, or even years.

3. Buy a piece of land and pave it with several inches of concrete, this being cheaper than long-run guard service.

To avoid such situations, not only must important radiation processes be carefully designed, but in addition the fire service should know in advance what tactics to use for particular situations. For example:

1. *Radioactive Gases.* Are they lighter or heavier than air? If heavier, what is the likelihood of the gas flow settling in basements or other occupied areas? Are the gases flammable or inert?

2. *Radioactive Liquids.* Are they lighter or heavier than water? What is the possibility of vapors being picked up by building ventilating systems or by stray air currents? Is there any likelihood of vapors or liq-

uids getting into sewers? Into waterways? Into occupied areas? Are the liquids flammable or inert?

3. *Radioactive Solids.* What is the particle size? Can the material become airborne or waterborne? Is the material flammable or inert?

4. *All Materials.* What is the radiological half-life? The biological half-life?

One very important procedure, sometimes overlooked, can be of tremendous value in assuring the proper handling of radiation fires. That procedure is making certain that there is close and regular coordination between fire prevention bureaus and the firefighting forces. Properly made, visits and inspections are valuable tools both for preplanning and for establishing firefighting procedures. Some of the items discussed in the next few paragraphs have been briefly mentioned in earlier chapters, but their importance to a well-coordinated program warrants re-emphasis.

Visits to plants or institutions for the purpose of discussing radiation hazard problems should not be made in a routine manner. Neither should these visits be made by regular inspectors. They should be made by qualified officers able to discuss the problem on equal terms with management. Their purpose is not to check for code violations (leave that to the regular inspectors). Rather, their purpose is to determine along with management what needs to be done in the event of fire that could occur long before code violations could be corrected. Management needs to be made fully aware of what the firefighting force needs in the way of information and assistance and why. Suggestions for improvements are of course in order, but the main purpose is to gain information and agreements for cooperation in the event the department had to respond to a fire that night. After each visit, information gained should be immediately incorporated into the firefighting procedures already developed for the plant in question.

The radiation hazards of the Atomic Age vary so widely that it would be impossible for any fire officer, even one thoroughly schooled in radiation problems, to assess a situation accurately if he arrived at 2:00 A.M. in front of a plant about which he knew nothing and found the remains of a radiation sign on the front door. Only two options exist for this situation and neither one is acceptable:

1. Fight the fire in a more or less conventional manner while awaiting the arrival of personnel with needed information and monitoring equipment. This course of action would expose personnel to quite possibly dangerous levels of radiation and result in unnecessary contamination of property and equipment.

2. Stand by and let the building burn while awaiting the arrival of knowledgeable individuals. This method assures that there will be large property losses and serious contamination.

While all fire activities require a degree of coordination among interested parties, radiation fires are in a class by themselves. Successful firefighting in this area with minimum danger to life and minimum property losses requires a high degree of coordination and cooperation among fire service organizations, fire prevention bureaus, radiological officers, and management. The need for free interchange of information can perhaps best be illustrated by a series of questions. Answers to these questions are of interest to all concerned. Rightly used, the answers point the way to correct fireground actions and to plant improvement programs.

1. Are radiation warning signs made of combustible or light noncombustible material that may be destroyed or made illegible by fire or heat prior to the arrival of

the fire department? If so, what alternate methods are in use to identify hazard areas and the type and extent of the hazard?

2. What is the radiation intensity from each source, shielded and unshielded?

3. Where and how are the materials stored?

4. What types of radiation are emitted?

5. Is the radioactive material in an area that is susceptible to a sustained fire that might destroy or damage the shielding?

6. What is the prefire condition of any shielding? (Do not overlook cracked or spalled concrete or the presence of temporary, substandard shielding.)

7. Are the types of containers used for radioactive material storage likely to be affected by fire or heat?

8. What are the physical and chemical properties of the radioactive materials insofar as they might affect firefighting procedures? Will they burn? Will they react chemically with other materials, including fire extinguishing agents, that may be present in the fire area? Can materials carrying radiation become airborne or waterborne?

9. What are the requirements for self-contained breathing apparatus?

10. Is prefire contamination control practice necessary for a potential fire situation? If so, what specific procedures are required to handle the problem with minimum risk?

11. Is radioactive material left out of its shield during nonworking hours? If so, have adequate provisions been made to notify the fire service and other safety organizations that might be affected?

12. Are radiation level checks necessary before entering the building? If so, what instrumentation is needed?

13. Assuming monitoring is required, does the fire service have the proper instruments and qualified people to operate them? If not, what provision has been made to correct the deficiency?

14. Are qualified plant personnel on duty at all times? If not, how is this lack of advisory personnel to be rectified?

15. Has each member of the firefighting force been assigned a personal dosimeter and/or film badge?

16. Is a current list maintained of agencies and individuals who are readily available for radiation emergencies?

17. Are fire and police departments notified in advance of hazardous shipments of radioactive materials to go through their areas? Are these shipments accompanied by couriers? If not, what provisions have been made for police escorts?

18. Are fire and police personnel familiar with the meanings of the different types of radiation warning placards and labels?

19. Is there a recognition program for the principal kinds of packaging used to transport radioactive materials?

20. Are transit shed operators aware of the regulations governing the number and types of radioactive materials packages that may be stored in one area?

21. Does the fire service maintain records of each individual's radiation exposure?

Assume that these questions and all others deemed

necessary have been asked and satisfactorily answered. Assume further that all these data have been integrated into the prefire planning program. By itself, this raw mass of data and plans does not help the firefighter too much. For example, there might be ten significant radiation locations which together supply 2,000 bits of information. To be usable at the fireground this information must be simplified. Fire officers who may need to make split-second decisions do not have time to review pages of data. They need and should have easily read running cards and charts. In the author's opinion there should be two running cards for each radiation location: (1) firefighting procedures and (2) decontamination requirements after the fire. The second card could be more detailed, as more time would be available for making decisions.

To complete the firefighting picture, fireground officers should be in two-way radio communication with headquarters. Personnel at headquarters can do much to assist the force at the fire. They have the facilities and the time to contact outside sources of assistance, arrange for relief of fireground personnel, expedite the delivery of supplies, provide advice based on a review of complete data plus information from the fireground, and so on.

The fire is out, the engines are back in the station, the cancer patient is getting cobalt treatment, food is being irradiated, reactors are humming, and what have we learned? Hopefully, we have learned through study and experience that while radiation is a hazard, it is a hazard that can be lived with safely. We should also be aware of the tremendous benefits for humanity that are available from the proper use of radioactive materials. Since it will not go away and is too useful to throw away, we should learn to live with it.

Do not ignore radiation. Respect but do not fear it. Understand and use it.

Appendix A

Glossary of Common Radiation Terms

ACCELERATOR. A machine or device used to accelerate electrically charged particles to very high speeds. These fast-moving particles can penetrate matter and are aimed at specific targets. Accelerators are used for: (1) studying atomic disintegration and transmutation, (2) production of certain radioisotopes, (3) some types of medical therapy, radiography, and (4) sterilization.

ACTIVATION. The process of making a material artificially radioactive. Activation may be accomplished in an accelerator or by neutron bombardment.

ALPHA PARTICLE. A small particle made up of two protons and two neutrons (identical with the nucleus of the helium atom) and having a positive electrical charge that is twice as great as the negative charge of an electron.

ATOM. The basic building block of nature. Each atom consists of a nucleus about which one or more electrons revolve. The nucleus consists of one or more protons and from no neutrons to many. The number of protons, electrons, and neutrons determines the nature of the material. Atoms can be split with the release of energy and can be converted from one element to another. These facts make possible the development of nuclear power and many industrial and medical processes that would be impossible without radioactive materials.

ATOMIC ENERGY. Energy released in nuclear reactions or disintegrations. Of primary importance is the energy released by fission (when a neutron splits an atom's nucleus into smaller pieces) or generated by fusion (when two nuclei are joined together under millions of degrees of heat).

ATOMIC NUMBER. The number of protons (positively charged particles) in the nucleus of an atom. All elements have different atomic numbers. For example, the atomic number of hydrogen is 1; that of carbon is 6; that of oxygen is 8; that of iron, 26; that of lead, 82; that of radium, 88; and that of uranium, 92.

ATOMIC WEIGHT. The weight of an atom relative to the weight of a hydrogen atom is 1. This weight is approximately equal to the sum of the number of protons and neutrons in the nucleus. For example, the atomic weight of helium is 4 (2 protons, 2 neutrons); that of lithium is 7 (3 protons, 4 neutrons); and that of aluminum is 27 (13 protons, 14 neutrons). While most atoms of any element have a standard number of protons and neutrons, many elements have a small percentage of atoms with one or more extra neutrons. These atoms will all have the same chemical properties, but the extra neutrons will modify some physical properties and will also slightly modify the atomic weight. Electrons for all practical purposes have no effect on atomic weight.

ATOM SMASHER. A machine (accelerator) that speeds up atomic and/or subatomic particles so that they become projectiles capable of breaking apart the nuclei of other atoms.

AUTORADIOGRAPHY. A process in which radioactive material is placed close to photographic film. The radiations "fog" the film, leaving an image or self-portrait of the source.

BACKGROUND RADIATION. Radiation coming from sources other than the radioactive material being checked. Background radiation is primarily from cosmic rays from outer space, although it may also be in part from other radiation sources in the environment. This radiation can be detected by a counter.

BETA PARTICLE. A small negatively charged particle thrown out by many radioactive materials. A beta particle is identical with an electron. (Beta radiation.)

BINDING ENERGY. The energy that holds the protons and neutrons of a nucleus together.

BIOLOGIC SHIELD. A shield provided to absorb neutrons that will inevitably escape from a reactor while fission is taking place. The shield is usually several feet of a dense concrete or its equivalent in other materials. Such shielding is essential for the protection of personnel.

BOMBARDMENT. The shooting of neutrons, alpha particles, and other high-energy particles at atomic nuclei, usually for the purpose of splitting the nuclei or forming new elements or variations of elements.

BREEDER REACTOR. A reactor that produces more nuclear fuel than it consumes. For example, a nonfissionable isotope bombarded by neutrons is converted to a fissionable material such as plutonium. These new fissionable materials are then usable as fuel.

CHAIN REACTION. When the nucleus of a fissionable atom is split by a neutron, energy and one or more neutrons are released. The free neutrons that strike other fissionable nuclei in turn release more energy and more neutrons. For this reaction to be self-sustaining a definite minimum amount of nuclear fuel must be present. *See* CRITICAL MASS.

CHARGE. The fissionable material used for fuel in a nuclear reactor. (Produces a chain reaction.)

COBALT 60. One of the radioactive isotopes of the element cobalt; a widely used, very important source of gamma radiation.

COFFIN. A thick-walled container, usually lead-lined, used for transporting radioactive materials.

CONTROL ROD. A device used to control the power output of a nuclear reactor. Materials used in control rods absorb free neutrons which otherwise would be splitting more fuel atoms. Pushing control rods in slows down the rate of atomic power production. Pulling the rods out increases the power output.

CONVERTER. A reactor that uses one type of fuel and produces a different fuel. For example, a reactor burning or using uranium 235 could simultaneously be used to produce plutonium.

CORE. The center of a nuclear reactor, where nuclei are split and power is produced.

COUNTER. A device used to measure radioactivity by counting nuclear disintegrations.

CRITICAL MASS. The specific minimum amount of nuclear fuel that must be present before any chain reaction can take place. Critical mass requirements are affected by the geometry of the mass and the presence or absence of moderators and/or reflectors. If too little fuel is present, more free neutrons will leave the mass than are produced by the splitting of nuclei and the reaction will die out.

CURIE. A measure of the rate at which a radioactive material throws off particles (or disintegrates). It is the radioactivity of 1 gram of radium (10 grams of radium would be a 10-curie unit).

DECAY. The process by which a radioactive material disintegrates to form another element. For example, a polonium atom disintegrates or decays to form lead. Alpha particles are ejected during this process.

DEUTERIUM. A heavy isotope of hydrogen. The nucleus contains 1 neutron in addition to the single proton. Each atom has twice the weight of a standard hydrogen atom.

DOSIMETER. An instrument used to determine the amount of radiation dosage an exposed person has received.

ELECTRON. An extremely small, negatively charged particle. Electrons rotate in orbits around the atom nucleus. The chemistry of an atom is governed by the number and arrangement of the electrons.

ELEMENT. A basic substance made up of a related group of naturally occurring isotopes. All atoms of an element have the same number of protons, but the number of neutrons can vary. Carbon, oxygen, iron, and tin are examples of elements.

FERTILE MATERIAL. A nonfissionable substance that can be converted to a fissionable material by reactor processes. For example, uranium 238 can be converted to plutonium. U238 is not a fissionable material, whereas plutonium is.

FILM BADGE. A special arrangement of photographic film worn by nuclear workers. Radiation exposure is checked by inspecting the film for the effects of radiation.

FISSILE CLASS. A classification of fissionable materials according to the type of controls required to assure nuclear criticality safety. Shipment of fissile materials, defined as plutonium (238, 239, and 241) and uranium (233 or 235) is regulated by the Department of Transportation, and materials fall into three classes: Fissile Class I, Class II, and Class III.

FISSION. The process of splitting an atomic nucleus into two parts with the evolution of large amounts of radioactivity and heat. Fission reactions occur only with certain heavy elements such as plutonium and uranium.

FISSIONABLE MATERIAL. An atomic nucleus which undergoes fission when bombarded by neutrons.

FUSION. The joining together of the nuclei of two light atoms to form a heavier nucleus. Fusion is accompanied by the release of a great deal of energy. The process is accomplished under conditions of extreme heat.

GAMMA RAYS (RADIATION). Long-range, highly penetrating rays of pure energy.
GEIGER COUNTER. An electrical device that detects radioactivity by counting the formation of ions.
HALF-LIFE. The time required for a radioactive isotope to lose half its strength (radioactive intensity). Half-lives may vary from a few seconds to billions of years.
HEAVY WATER. Water containing deuterium (a heavy hydrogen isotope) instead of ordinary hydrogen. Heavy water is used in many reactors to slow down neutrons.
HOT. A term used colloquially to mean highly radioactive.
ION. An atom which has gained one or more electrons to have a negative charge or has lost one or more electrons to have positive charge.
IONIZATION CHAMBER. A device used to measure radioactivity.
ISOTOPE. One of two or more nuclei of the same element that have the same electrical charge but different masses. Isotopes have the same number of protons but a different number of neutrons. For example, both uranium 235 and uranium 238 have 92 protons per atom. U238, however, has 146 neutrons per atom while U235 has only 143. The atomic weight of a U238 atom is therefore 3 higher than that of a U235 atom.
KILOCURIE. A unit (1,000 curies) used to measure radioactivity.
LINEAR ACCELERATOR. A machine designed to speed up charged particles. These particles move in a straight line, not in the spiral and curved paths used in most types of accelerators.
MILLIROENTGEN. One-thousandth of a roentgen; a unit for measuring radioactive doses.
MODERATOR. A material used to moderate or slow down neutrons in a reactor. Slowed-down neutrons are more effective than fast neutrons in promoting fission and in maintaining desired chain reactions.

MONITOR. A detector used to determine whether or not the radiation level in an area is safe for workers.
NEUTRON. A basic atomic particle consisting of a proton and an electron bound together by internal energy. It has no electrical charge and therefore can readily penetrate a positively charged nucleus. When a neutron loses its electron, it becomes a proton.
NUCLEAR BOMBARDMENT. Shooting atomic projectiles at the nuclei of atoms in order to split the atom or to form new elements.
NUCLEAR ENERGY. The energy released by a nuclear reaction such as fission or fusion.
NUCLEAR REACTION. As commonly used, the result of the bombardment of nuclei by atomic or subatomic particles. The term, however, also refers to the decay of radioactive material.
NUCLEUS. The central core of an atom, consisting of protons and neutrons locked together.
PLUTONIUM. One of the heavy elements that fissions when bombarded by neutrons. Plutonium is a useful fuel for nuclear reactors and can also be produced in "breeder" reactors. For all practical purposes plutonium is a man-made element.
PROTON. One of the basic particles of a nucleus. Its charge is identical in amount to that of the electron, but is positive.
RAD. The standard unit for measuring the absorbed dose of any type of radiation. The effects of an absorbed dose of x rads from one type of radiation can be quite different from the effects of x rads from another type. These differences are expressed by RBE (relative biological effectiveness).
RADIOACTIVITY (RADIATION). The emission of very fast atomic particles or rays from atoms. Major forms of radiation are alpha, beta, and gamma. Some elements are naturally radioactive while others are made radioactive by being bombarded by neutrons or other particles.

RADIOISOTOPE. An isotope of an element that is radioactive. Some radioisotopes occur naturally, but the majority are produced in nuclear reactors by bombarding the material with neutrons. Their use is widespread throughout industry and in scientific research.

REACTOR. An atomic furnace in which the nuclei of the fuel elements undergo fission when bombarded by neutrons. New neutrons are produced by the fission process and a self-sustaining chain reaction started. This process releases large amounts of energy, which in turn generates large amounts of heat. In power reactors the heat (through heat transfer mediums) is used to generate power. Some small research reactors are, however, arranged to dissipate the heat, and these produce no power.

REFLECTORS. Layers or blankets of materials used in some reactors to surround fuel cores to reduce the escape of neutrons.

RELATIVE BIOLOGICAL EFFECTIVENESS (RBE). A measure of the relative effects on the body of different types of radiation from different types of materials.

REM. A measure of the effect of radiation on the body (used for all types of radiation). This term should not be confused with the term *rad*.

ROENTGEN. The unit used for measuring penetrating external radiation exposure.

SHIELD. A wall or barrier which protects workers from radiation hazards.

SOURCE MATERIAL. Any substance that emits radiation.

THERMONUCLEAR REACTION. Fusion; the combining of two light nuclei to form a heavy atom with the release of a large amount of energy.

THORIUM. A heavy element that on bombardment by neutrons becomes fissionable uranium.

TRACER. A radioisotope that has been mixed with a stable material so the material can be traced as it undergoes physical and chemical changes. This property is of great value in many phases of science, industry, and agriculture.

TRANSMUTATION. The process by which one element is changed into another. For transmutation to occur it is necessary for the number of protons in the nucleus to be changed.

TRITIUM. The radioactive isotope of hydrogen. The atom nucleus contains one proton and two neutrons and is three times as heavy as an atom of ordinary hydrogen.

UNSTABLE ELEMENT. Any radioactive element (all of these emit particles and decay to form other elements).

URANIUM. One of the heavy metals and the only one which as found in nature contains appreciable amounts of isotopes with fissionable nuclei.

X-RAY. Highly penetrating radiation that is quite similar to gamma rays but does not come from the nucleus but from electrons surrounding the nucleus. X-rays are produced by electron bombardment.

Appendix B

Table A
Radiation Characteristics of Common Isotopes

Isotope	Radiation		Half-Life			Quantity of Material for One Curie
	Beta	Gamma	Years	Days	Hours	
Antimony 122	*	*			66	
Antimony 125	*	–	2.78			
Arsenic 77	*	*			38.7	
Barium 140	*	*		12.8		
Bismuth 210	*	–		5		
Bromine 82	*	*			35.6	
Carbon 14	*	–	5.57 $\times 10^3$			0.23 Grams
Cesium 137	*	–	30			11.5 Milligrams
Chromium 51	–	*		27		1.1 Micrograms
Cobalt 60	*	*	5.24			880 Micrograms
Gold 198	*	*			64.8	4.1 Micrograms
Hafnium 181	*	*		44.6		
Hydrogen 3	*	–	12.5			0.1 Milligrams
Iodine 130	*	*			12.5	
Iodine 131	*	*		8.05		8.2 Micrograms
Iridium 192	*	–		74.2		11 Micrograms
Iron 59	*	*		44.3		20.4 Micrograms
Krypton 85	*	*	10.3			2.5 Milligrams
Mercury 203	*	*		45.4		
Nickel 63	*	–	125			
Osmium 191	*	*		16		
Phosphorus 32	*	–		14.3		3.5 Micrograms
Potassium 42	*	*			12.5	0.16 Micrograms
Rubidium 86	*	*		18.7		
Ruthenium 106	*	–	371			
Sodium 24	*	*			15.1	0.12 Micrograms
Strontium 89	*	–		50.5		36 Micrograms
Strontium 90	*	–	28			63 Micrograms
Sulfur 35	*	–		89		
Tungsten 185	*	*		75.8		
Yttrium 90	*	–			64.4	
Yttrium 91	*	*		59.1		
Zirconium 95	*	*		65		

* = Yes.
– = No.

Table B
Radiation Characteristics of Some Heavy Elements

Radioactive Material	Radiation Beta	Radiation Gamma	Half-Life Years	Half-Life Days	Quantity of Material for One Curie
Radium 226	Alpha	*	1,620		1 Gram#
Plutonium 239	Alpha	*	24,000		160 Milligrams§
Polonium 210	Alpha	*		138	223 Micrograms
Thorium 232	*	—	$1.4x\ 10^{10}$		9.07 Metric Tons
Uranium 235	Alpha	*	$7.1x\ 10^{8}$		476 Kilograms§
Uranium 238	Alpha	—	$4.5x\ 10^{9}$		2.99 Metric Tons

§ = Fissionable Material
# As radium disintegrates, offspring give off gamma rays
Note that many heavy elements give off alpha radiation

Table C
Examples of Gamma Radiation Levels from 1 curie of Selected Radioisotopes at Various Distances.

Material	Roentgens per hour At 1 ft.	At 3 ft.	At 6 ft.
Cesium 137	3.87	0.43	0.11
Cobalt 60	14.31	1.59	0.40
Gold 198	2.70	0.30	0.08
Iodine 131	2.52	0.28	0.07
Sodium 24	20.79	2.31	0.58
Zinc 65	3.24	0.36	0.09

Resources

Department of Transportation
400 7th St., N.W.
Washington, D.C. 20590

A Review of the Department of Transportation (DOT) Regulations for Transportation of Radioactive Materials.

U.S. Atomic Energy Commission*

Operational Accidents and Radiation Exposure Experience, 1943–1970.
Your Body and Radiation (N. A. Frigerio).
Radiation Hazards in Firefighting.

U.S. Energy Research and Development Administration
20 Massachusetts Ave., N.W.
Washington, D.C. 20545

Living with Radiation, Vol. 1 and 2 (F. L. Brannigan).
The Genetic Effects of Radiation (I. Asimov and T. Dobzhansky).

---

*AEC has now been replaced by other Federal agencies. These publications will eventually be replaced. For the present they should be available from the Government Printing Office, North Capitol and H Sts., N.W., Washington, D.C. 20401

Emergency Handling of Radiation Accident Cases (Police).

National Fire Protection Association
470 Atlantic Ave.
Boston, Mass. 02210

Fire Officer's Guide to Dangerous Chemicals (C. W. Bahme).
Fire Protection Handbook.
Radiation Control (A. A. Keil).
Standard 801 Facilities Handling Radioactive Materials.
Standard 802 Fire Protection Practice for Nuclear Reactors.

Factory Mutual System
Factory Mutual Engineering Corp.
Factory Mutual System
1151 Boston–Providence Turnpike
Norwood, Mass. 02062

Handbook of Industrial Loss Prevention

Department of the Army
The Pentagon 20310 or
Office, Chief of Engineers
Forrestal Bldg.
Washington, D.C. 20560

TB 385-2 *Nuclear Weapons, Firefighting Procedures.*

Conversion Table

1 inch = 2.54 centimeters
1 foot = 0.3048 meter
1 yard =0.9144 meter
1 mile =1.609 kilometers
Temperature Centigrade = 5/9(Temperature Fahrenheit − 32)

Index